N.R. Mandal

D1760864

Aluminum Welding

WOODHEAD PUBLISHING LIMITED

Cambridge England

ALPHA SCIENCE INTERNATIONAL LTD.

Pangbourne

Dr. Nisith R. Mandal
Associate Professor
Department of Ocean Engineering & Naval Architecture
Indian Institute of Technology, Kharagpur-721 302, India

Copyright © 2002

Alpha Science International Ltd
P.O. Box 4067, Pangbourne RG8 8UT, UK

This book is to be sold in UK and Europe only by
Woodhead Publishing Limited, Abington Hall,
Abington, Cambridge CB1 6AH, England

British Library Cataloguing in Publication Data
A catalogue record for this book is available from the British Library.

Woodhead Publishing   ISBN 1-85573-597-0
Alpha Science   ISBN 1-84265-095-5

Printed in India.

£45.00   S.I.

671.52

14664L

Aluminum Welding

Preface

Unlike steel welding, aluminum welding is far more complicated because of the inherent thermo-physical properties of aluminum. Hence to achieve quality and class approved welding in aluminum, stringent process control is necessary. This can be achieved only through proper and adequate knowledge of the material as well as all the process variables.

While teaching 'Ship Construction & Welding Technology' and 'Marine Construction & Repair Techniques' as regular courses in undergraduate and graduate curricula and various short-term courses and projects carried out by me during the past 15 years, I came across a couple of books dealing with welding in general. However, I could not find any book, except American Welding Society Handbook, which dealt with aluminum welding. This inspired me to get on this job and provide practicing engineers and students of welding technology with a comprehensive book with logically organized contents spread over several chapters.

This book contains an elaborate introductory description of the properties of wrought and cast aluminum alloys and the variety of their applications. There is an International Accord recognizing the Aluminum Association Wrought Alloy Designation System as the *de facto* international standard. The alloy designation system, first for wrought alloys and then for cast alloys is followed by a brief description of the temper designation system. Composition, properties and typical application of nonheat-treatable and heat-treatable aluminum alloys and the effect of alloying elements on the mechanical properties of the alloys are discussed in detail. The principal characteristics and applications of various aluminum alloys and cast alloys with examples have been included. As the selection of the correct filler metal greatly influences the service life of an aluminum weldment, the chemical composition of aluminum filler metals and a guide to selection of filler metals for various aluminum alloy combinations form a part of the text.

Edge preparation and proper assembly is very important in aluminum welding. The extra time needed to ensure a close fit is often less than the extra time required in welding an improperly prepared assembly. The effect of fixtures and preheating requirement has been covered in this work.

Arc welding power sources can be as varied as the welding processes. To meet the unique electrical requirements of various arc welding processes many types of power sources are necessary. Different types of power sources and their effect in aluminum welding has been elaborately dealt with.

Weld quality and weld deposition rate, both are influenced very much by various welding parameters, shielding gases and joint geometry. Each of these parameters has influence to a varying degree on the deposition rate, weld bead shape, depth of penetration, cooling rate and weld induced distortion. Hence to achieve a sound welded joint with adequate metal deposition rate and minimum weld induced distortion, a proper understanding of the effects of these parameters or process variables is needed. These aspects have been addressed in detail.

The weld metal transfer mechanism and its dependence on process variables in case of aluminum welding has been explained in a simple manner, such that a wide spectrum of reader can understand this complicated phenomenon.

Electrodes for gas tungsten arc welding, electrode sizes, electrode tip configuration, shielding gases, and the process variables, all these interact with each other very strongly, therefore they can not be treated as independent variables while establishing welding procedures for fabricating specific joints. General problems encountered in gas tungsten arc welding of aluminum alloys and their causes are mentioned in detail.

The distortion mechanism, its type and causes have been explained in details. Basics of numerical methods for analysing thermal distortion and formation of residual stresses have been incorporated. To effectively analyze the problem and account for the nonlinear phenomena associated with it, one has to draw upon knowledge from a variety of scientific disciplines, including heat transfer, applied mechanics, numerical analysis, and material science. The practicing engineer would like to have a design tool to account for residual stresses and their effects. The physical basis of thermal distortion and the factors affecting it are not difficult to understand, and therefore empirically determined formulae are adequate for practical purposes. It is not the intent of the author to present in detail all the mathematics involved in the discussed methodologies. The emphasis will rather be placed on how the designers and fabricators can effectively use the state-of-the-art methodologies available today for solving the residual stress problem.

N.R. MANDAL

Acknowledgements

The author wishes to acknowledge the contributions of the officers and engineers of various shipyards and the officers of Warship Overseeing Team, for their active participation in the discussions related to weld induced distortions during the various short courses conducted by the author. The author is also grateful to the management of Garden Reach Shipbuilders and Engineers Ltd., Kolkata and Mazagon Dock Ltd., Mumbai, for their help to carry out real life studies and full scale tests in their respective shipyards.

The author wishes to thank Shri Biswajit Chatterjee and Shri Ranjit Banik for their help in preparing the manuscript and autocad drawings. The support provided by Continuing Education Programme of Indian Institute of Technology, Kharagpur, is acknowledged. This work would not have been completed without the support of my wife, Suparna, who has been considerate enough during those long hours of my involvement with this work over the years.

Contents

1

General Characteristics of Aluminum

Aluminum is light in weight, yet some of its alloys have strengths exceeding the mild steel. It retains good ductility at subzero temperatures, has high resistance to corrosion, and is not toxic. Pure aluminum melts at 660°C. Aluminum alloys have an approximate melting range from 482 to 660°C, depending upon the alloy. There is no color change in aluminum when heated to the welding temperature range. High thermal conductivity (as compared to steel) necessitates a high rate of heat input for fusion welding. Thick sections may require preheating. When resistance spot welding, aluminum's high thermal and electrical conductivity require higher current, shorter weld time, and more precise control of the welding variables than when welding steel. Aluminum and its alloys rapidly develop a tenacious, refractory oxide film when exposed to air.

What Makes Aluminum Different?

The main characteristics of aluminum, which influence welding, are hydrogen solubility, aluminum oxides, thermal conductivity, thermal expansion and the solidification shrinkage, and non-discolouration. It is a lack of understanding in the cleaning aspect that has kept many fabricators away from the welding of aluminum.

One needs to understand that moisture or hydrocarbons on the surface might decompose beneath the arc, and create hydrogen. Hydrogen is extremely soluble in molten aluminum, so, when the weld freezes, the hydrogen can get trapped in the weld itself. That is the main source of porosity in aluminum welds. From the standpoint of cleanliness, any source of moisture have to be removed from the surface. Once this is done, the main source of porosity in aluminum welds is eliminated. The cleanliness of the filler metal is another consideration. The importance of the storage and handling of the filler metal is often underestimated. The filler metal has to be handled just like the plate, and for the same reasons. Most aluminum wire feeders have an enclosure for wire spool to keep the normal contaminants off the wire, although that does not necessarily address the oxides.

It is much like the hermetically sealed filler metal used to weld HSIA steels. Aluminum filler metal has to be kept dry and free of lubricants.

The oxides, in a sense, can be even more of a problem. Aluminum oxide melts at about 2066°C or about three times the melting point of the aluminum alloy itself. It is obvious in welding that the base metal will be melted long before the oxide. In the welding of nonheat-treatable aluminum alloys, the natural oxides can be broken up by the inert arc. However, in welding heat-treatable alloys that have been taken up to a high temperature to give them a certain thermal treatment, a much thicker oxide will form than is present in the nonheat-treatable alloys.

Aluminum oxide is an insulator. It does all kinds of wild and crazy things to the arc. It can cause preferential arcing or, if thick enough, no arcing at all. It is the major cause of poor fusion in aluminum welds. The stainless brush is probably the best way to preclean after a thorough degreasing. There are all kinds of chemical solutions that will remove the oxides, but they involve multistep operations. The need for removing the oxide is to provide uniform melting and proper fusion. In the 5000 series, the oxide is most hygroscopic and is most sensitive to moist conditions. In that state, if not removed, it will contribute to the formation of porosity in the welds.

High Thermal Conductivity

The thermal conductivity of aluminum is about six times that of steel. If aluminum is welded too slowly, the heat travels ahead of the arc and will force the operator to make continual adjustments to current and travel speed. Even though most aluminum alloys melt in the 565° to 648°C range, the high thermal conductivity of aluminum necessitates that a more intense heat be employed than that used in welding steel. Welding is carried out with higher heat input and at a faster rate than that used for steel. Conductivity is also important when weld procedures are being set and welders qualification is being carried out. It can affect the mechanical properties. Larger or longer test pieces are to be used in order to provide an adequate heat sink or avoid overheating.

Thermal Expansion

Thermal expansion is about twice that of steel. Of particular concern here is the greater expansion in thinner materials. Solidification shrinkage in aluminum weld metal is about 6% by volume and it can be the main cause for distortion, especially in thicker welds. To compensate, the weld passes have to be balanced, particularly in fillet welds.

Unlike steel, aluminum does not change colour when it is heated up. In welding the metal, the operator has to get in position where he can see the molten pool and the end of the electrode tip if he expects to make good welds. Aluminum surfaces are also highly reflective. As such, the welders require protection against radiation.

Dilution is a Factor

A good filler alloy dilution into the weld joint is needed. Interpass cooling should be used to keep peak temperatures as low as possible during welding. Chill bars and water-cooled backings can be used to take the heat away. The wide melting range and the use of dissimilar metal fillers on the heat-treatable alloys can make them somewhat more sensitive to weld cracking. Filler Alloy 4043, commonly used to weld 6061 aluminum, is a nonheat-treatable alloy. Magnesium dilution from the 6061 is required to provide response to postweld heat treatment. To avoid this need for base metal dilution, a 4643 filler metal is used. The advantage with the heat-treatable alloys is, higher as-welded strengths can be developed than can be achieved with the nonheat-treatable grades. The reason is that postweld thermal treatments can be applied. However, there will be a loss in ductility. For example, the non heat-treatable 5456 grade, as welded, provides high ductility and toughness, with a minimum tensile strength of about 289 MPa. That is about the same as the minimum tensile strength of 6061-T6 aluminum prior to welding. However, the nonheat-treatable alloys have lower yield strengths. With 5456, the yield strength will be 172 MPa after welding. On the other hand, if Alloy 6061 is reheat-treated, the yield strength will go up to 241 MPa.

Alloys 7004, 7005 and 7039 are the least sensitive to rate of quench. From the heat of welding, they quench themselves. These alloys naturally age in air. The day 7005 is welded, one would get about 234 MPa tensile strength across the weld. A week later, it would be up to about 289 MPa. In about two weeks to a month, it will be up to around 344 MPa. Artificial aging is needed in order to increase the yield strength substantially after welding. These 7000 series alloys will develop a high as-welded strength in time without postweld thermal treatment.

The Effect of Power Supplies

When the GMAW process was first used, constant current power supplies or those having drooping volt-ampere characteristics were used for all metals. The torch contact tubes were 100 to 150 mm long, and a "touch" start or slow run-in wire feed feature was used to initiate the arc. This approach is still preferred for the highest quality, lowest maintenance GMA production welding of aluminum.

But, because of certain problems in the welding of steel, constant voltage (CV) power supplies were soon developed and the contact tubes were shortened to an inch or less. One problem there had been the resistance heating effect that took place when electrode extension was used to increase weld deposition rates. Also, since the CV power supplies could produce high current surges at starts, there was no longer any need at least in the welding of steel, for touch starts and slow run-ins. As a result, both of those features were soon dropped from most suppliers' wire feed controls.

As might be expected, these decisions have turned out to be harmful for the welding of aluminum. Due to the insulating characteristic of aluminum oxide, the CV power source has introduced excessive arcing at the few points of current communication in the short contact tube during the high current surges at weld starts. After 30 to 50 arc starts, "burn-backs" can be expected to occur. By contrast, operators of Drooper power supplies can expect to have thousands of starts without any incidences of burn-back. Due to the fact, that longer contact tubes are available in a touch start system, there are more points of contact. As such, the current can be brought along gradually and there are no surges.

When using a CV power supply, torch manipulation in the root of a fillet or V-groove joint creates wide fluctuations in heat input in the thermally conductive aluminum. The result is discontinuities in linear root penetrations. A constant current power source will reduce these fluctuations considerably. Other advantages of the constant current power source with the "touch start" include the following:

- Improved weld fusion at weld starts.
- Lower weld profile at starts.
- Improved weld soundness when over-lapping tack welds or weld starts are used.
- The ability to "crater fill" weld terminations with several quick shots.

Only constant current power supplies are suitable for high current, high deposition GMA welding of thick aluminum plate in excess of 400 A. The new inverter power supplies may be a solution. These machines can furnish a constant current output at a constant voltage, the result being a very stable heat input to the arc.

1.1   Aluminum Alloy and Temper Designation System

There is an International Accord recognizing the Aluminum Association Wrought Alloy Designation System as the de facto international standard. This accord has been ratified by almost all of the world's aluminum producing countries. The alloy designation system is briefly described below, first for wrought alloys and then for cast alloys. These are followed by a brief description of the temper designation system. More detailed information on The Aluminum Association Alloy and Temper Designation System may be gained from study of section 1.0 of Aluminum Association publication "Aluminum Standards and Data"

1.1.1   Alloy Designation System—Wrought Alloys

Pure aluminum is readily alloyed with many other metals to produce a wide range of physical and mechanical properties. This means by which the alloying elements strengthen aluminum are used as the basis to classify aluminum alloys into two categories: nonheat-treatable and heat-treatable.

- First digit—Principal alloying constituent(s)
- Second dight—Variations of initial alloy
- Third and fourth digits—Individual alloy variations (number has no significance but is uniue)
  - 1xxx—Pure Al (99.00% or greater)
  - 2xxx—Al-Cu Alloys
  - 3xxx—Al-Mn Alloys
  - 4xxx—Al-Si Alloys
  - 5xxx—Al-Mg Alloys
  - 6xxx—Al-Mg-Si Alloys

- 7xxx—Al-Zn Alloys
- 8xxx—Al + Other Elements
- 9xxx—Unused Series

1.1.2   Alloy Designation System—Casting Alloys

- First digit—Principal alloying constituent(s)
- Second and third digits—Specific alloy designation (number has no significance but is unique)
- Fourth digit—Casting (0) or ingot (1, 2) designation
- Variations indicated by preceding letter (A, B, C)
  - 1xx.x—Pure Al (99.00% or greater)
  - 2xx.x—Al-Cu Alloys
  - 3xx.x—Al-Si + Cu and/or Mg
  - 4xx.x—Al-Si
  - 5xx.x—Al-Mg
  - 7xx.x—Al-Zn
  - 8xx.x—Al-Sn
  - 9xx.x—Al + Other Elements
  - 6xx.x—Unused Series

1.1.3   Temper Designation System

F → As fabricated
Applies to products of forming processes in which no special control over thermal or strain-hardening conditions is employed. Mechanical property limits are not assigned to wrought alloys in this temper but are assigned to cast alloys in the "as-case" (F) temper.

O → Annealed
Applies to wrought and cast products which have been heated to produce the lowest strength condition and to improve ductility and dimensional stability.

H → Strain-hardened
Applies to wrought products which are strengthened by strain-hardening through cold-working. The strain-hardening may be followed by supplementary thermal treatment, which produces some reduction in strength. The "H" is always followed by two or more digits.

W → Solution heat-treated
An unstable temper applicable only to alloys which age spontaneously at room temperature after solution heat-treatment. This designation is specific only when the period of natural aging is specified; for example, W1/2 h. Solution heat-treatment involves heating the alloy to approximately 538°C to bring the alloying elements into solid solution, followed by rapid quenching to achieve a supersaturated solution at room temperature.

T → Thermally treated to produce stable tempers other than F, O or H
Applies to products which have been heat-treated, sometimes with supplementary strain-hardening, to produce a stable temper. The T is always followed by one or more digits.

Subdivisions of T Temper: Thermally Treated
The first digit indicates specific sequence of treatments:

T1 → Naturally aged after cooling from an elevated temperature shaping process, such as extruding.

T2 → Cold worked after cooling from an elevated temperature shaping process and then naturally aged.

T3 → Solution heat treated and naturally aged.

T5 → Artificially aged after cooling from an elevated temperature shaping process.

T6 → Solution heat treated and artificially aged.

T7 → Solution heat treated and stabilized (overaged).

T8 → Solution heat treated, cold worked and artificially aged.

T9 → Solution heat treated, artificially aged and cold worked.

T10 → Cold worked after cooling from an elevated temperature shaping process and then artificially aged.

The second digit indicates variation in basic treatment.

Example: T42 or T62 = heat treated to temper by user.
Additional digits indicate stress relief. Examples: TX51 or TXX51 = stress relieved by stretching. Or, TX52 or TXX52 = stress relieved by compressing.

Subdivisions of H Temper: Strain Hardened
The first digit indicates basic operations:

H1 → Strain hardened only.

H2 → Strain hardened and partially annealed.

H3 → Strain hardened and stabilized.

The second digit indicates degree of strain hardening:

HX2 → Quarter hard.

HX4 → Half hard.

HX6 → Three-quarters hard.

HX8 → Full hard.

HX9 → Extra hard.

The third digit indicates a variation of a two-digit temper.

1.2 Nonheat-Treatable Aluminum Alloys

The initial strength of the nonheat-treatable aluminum alloys depends primarily upon the hardening effect of alloying elements such as silicon, iron, manganese and magnesium. These elements affect increase in strength either as dispersed phases or by solid-solution strengthening. The nonheat-treatable alloys are mainly found in the 1XXX, 3XXX, 4XXX, 5XXX alloys series depending upon their major alloying elements. Composition and typical application of nonheat-treatable aluminum alloys are shown in Table 1.1.

Table 1.1   Composition and applications of nonheat-treatable wrought alloys

Aluminum association designation	Nominal composition (% alloying element)				Typical applications
	Cu	Mn	Mg	Cr	
5083	–	0.7	4.4	0.15	Marine components, drilling rigs.
5086	–	0.45	4.0	0.15	Marine components, tanks, tankers, truck frames.
5466	–	0.8	5.1	0.12	Marine components.

Magnesium is the most effective solution-strengthening element in the nonheat-treatable alloys. Aluminum-magnesium alloys of the 5XXX series have relative high strength in the annealed condition. The strength of all the nonheat-treatable alloys may be improved by strain hardening. When fusion welded, the nonheat-treatable alloys lose the effects of strain hardening in a narrow heat-affected zone (HAZ) adjacent to the weld; the strength in the HAZ will approach that of the annealed condition. To remove the effects of strain hardening and improve ductility, the nonheat-treatable alloys may be annealed by heating to a uniform temperature in the range of 343 to 410°C. Although the rate of cooling from the annealing temperature is unimportant, fixturing may be required to prevent distortion or warping. The 5XXX series aluminum alloys are based on magnesium; they have the highest strengths of the nonheat-treatable alloys. They are typically available as sheet and plate product forms. All of these alloys are fusion weldable with a minimum loss of strength and those possessing more than 3.5% Mg are essentially free from hot cracking. The applications are numerous in the transportation fields. They can be resistance spot welded, but, because of their heavy oxide films, extra care in surface preparation is necessary.

1.3   Heat-Treatable Aluminum Alloys

The initial strength of aluminum alloys in this group depends upon the alloy composition, just as the nonheat-treatable alloys. Heat-treatable aluminum alloys develop their properties by solution heat treating and quenching, followed by either natural or artificial aging. Cold working may add additional strength. The heat-treatable alloys may also be annealed to attain maximum ductility.

This treatment involves holding at an elevated temperature and controlled cooling to achieve maximum softening. The heat-treatable alloys are found primarily in the 2XXX, 6XXX and 7XXX alloy series (Table 1.2). Although the 6XXX series alloys are prone to hot cracking, this condition can be readily overcome by correct choice of joint design and electrode. Care in joint and component design is also needed because of losses in strength that are obtained in the heat-affected zones. These HAZ strengths can be improved, however, through postweld heat treatment. These alloys can be soldered, brazed and resistance spot welded satisfactorily.

Table 1.2 Composition of Heat-Treatable 6061 Aluminum Alloy

Aluminum association designation	Nominal Composition (% alloying element)							Typical applications
	Cu	Si	Mn	Mg	Zn	Ni	Cr	
6061	0.25	0.6	–	1.0	–	–	0.20	Marine application

1.4 Filler Metal

The end use of the weldment and desired performance are important considerations in selecting an aluminum alloy filler metal. Many base-metal alloys and alloy combinations can be joined using any one of several filler metals, but only one may be the optimum for a specific application.

The primary factors commonly considered when selecting an aluminum alloy filler metal are,

- Freedom from cracks
- Tensile or shear strength of the weld metal
- Weld ductility
- Service temperature
- Corrosion resistance
- Color match after anodizing

Cracking

In general the nonheat-treatable aluminum alloys can be welded with a filler metal of the same basic compositions as the base alloy. The heat-treatable alloys are somewhat more complex metallurgically and are more sensitive to "hot short" cracking during the weld cooling cycle. Generally, a dissimilar filler metal having a lower melting temperature and similar or lower strength than the base metal is used for the heat-treatable alloys. By allowing the low-melting constituents of the base metal adjacent to the weld to solidify before the weld metal, stresses are minimized in the base metal during cooling and intergranular cracking tendencies are minimized. High silicon content and the high magnesium content aluminum alloys are easy to weld due to low sensitivity to cracking. 6XXX series alloys are very sensitive to cracking if the weld metal composition remains close to the base metal composition. These can be welded easily if beveled to permit an excess of filler metal admixture with the base metal. For alloy 6061, the weld metal should possess at least 50% alloy 4043 filler metal or 70% alloy 5356 filler metal. Fillet welds permit this admixture by the filler metal naturally, provided the base metal is not excessively melted. Filler metals with a high silicon content (4XXX series) should not be used to weld high magnesium content 5XXX series alloys; excessive magnesium-silicide eutectics developed in the weld structure decrease ductility and increase crack sensitivity. Mixing the high-magnesium content and high-copper content alloys results in high sensitivity to weld cracking and low weld ductility.

Strength
In many cases several filler metals are available that meet the minimum as-welded mechanical properties. Typical aluminum filler metal properties in as-welded condition are given in Table 1.3. The diffusion of alloying elements from base metal may increase the as-welded mechanical properties. The minimum shear strength and tensile strength of welds are highly dependent upon the filler metal composition as shown in Table 1.3. Filler metal alloys 5356, 5183 and 5556 provide high shear strength for structural fillet welds. The 1XXX and 5XXX alloy series filler metals produce very ductile welds and are preferred when the weldment is subjected to forming operations or post-weld straightening operations.

Table 1.3   Typical aluminum filler-metal properties (as-welded condition)

Filler alloy	Minimum shear strength (MPa)	All-weld-metal ultimate tensile strength (Mpa)
4043	79	200
5183	128	283
5356	117	262
5554	117	230
5556	138	290
5654	83	221

Corrosion Resistance
Aluminum-magnesium filler metals are highly resistant to general corrosion when used with base alloys having similar magnesium content. However, the 5XXX alloy series filler metals can be anodic to the 1XXX, 3XXX and 6XXX alloy series base metals with which they might be used. In immersed service, the weld metal will pit and corrode, to protect the base metal, at varying rates based upon the difference in electrical potential of weld and base metal. Thus an aluminum-silicon filler metal, such as alloy 4043 or alloy 4047, would be preferred for improved corrosion resistance over alloy 5356 filler metal when welding alloy 6061 base metal for an immersed-service application.

Which Filler Metal to Use?
The role of the filler metal in the welding of aluminum is a critical one. The high magnesium content filler alloys (5356, 5183 and 5556) provide the highest as-welded strength, permit the smallest fillet welds, and often result in financial advantages. Either 4043 or 5356 can be used to weld 6061 aluminum. For a shear strength requirement of 70 kg per linear mm, a 6 mm fillet is adequate with

5356 filler alloy; however, a 9 mm fillet is required with 4043 filler. The latter would normally require three weld passes at triple the weld time needed, compared to the 5356 filler metal. The larger fillet can also result in increased weld metal shrinkage. As the temperature decreases, aluminum alloys and their weldments gain strength. The aluminum alloys do not become brittle but maintain or increase in ducility as the temperatures decrease below zero. The high magnesium content aluminum alloys, like 5083, have exhibited very good cryogenic properties, and, because of their excellent weldability, have been widely used for the storage of liquid oxygen and for LNG.

With the heat-treatable aluminum alloys, the joint bevel, particularly with the 6XXX series, should allow the weld deposit to consist of 50% or more of filler alloy. Alloy 6061, even in thicknesses down to 2.3 mm, is beveled to permit a minimum of 50% dilution in butt joints when using 4043 filler alloy.

Selection of Filler Metal
Common criteria to be considered in selecting a filler metal are ease of welding, strength, ductility, corrosion resistance of the filler metal/base metal combination, color match with the base metal after anodizing, and service at elevated temperature. The selection of the correct filler metal greatly influences the service life of an aluminum weldment. Chemical composition of aluminum filler metals are given in Table 1.4. A guide to selection of filler metals for various aluminum alloy combinations is given in Table 1.5.

Table 1.4   Chemical composition of wrought aluminum filler metals

Filler alloy	Elements, (Wt. %)							
	Si	Fe	Cu	Mn	Mg	Cr	Zn	Ti
4043	4.5–6.0	0.8	0.30	0.05	0.05	–	0.10	0.20
4047	11.0–13.0	0.8	0.30	0.15	0.10	–	0.20	–
5183	0.40	0.40	0.10	0.50–1.0	4.3–5.2	0.05–0.25	0.25	0.15
5356	0.25	0.40	0.10	0.05–0.20	4.5–5.5	0.05–0.20	0.10	0.06–0.20
5554	0.25	0.40	0.10	0.50–1.0	2.4–3.0	0.05–0.20	0.25	0.05–0.20
5556	0.25	0.40	0.10	0.50–1.0	4.7–5.5	0.05–0.20	0.25	0.05–0.20
5654	Note a	Note a	0.05	0.01	3.1–3.9	0.15–0.35	0.20	0.05–0.15

Note a: Silicon plus iron shall not exceed 0.45 percent.

Table 1.5   Guide to the selection of filler metal for general purpose welding

Base metal	6005, 6061, 6063, 6101, 6151, 6201, 6351, 6951	5086	5083
5083	ER5356	ER5356	ER5183

Storage and Use of Aluminum Filler Metal
To avoid contamination, filler metal supplies must be kept covered and stored in a dry place at a relatively uniform temperature. Electrode spools temporarily left

unused on the welding machine, as between work shifts, should be covered with a clean cloth or plastic bag if the feed unit does not have its own cover. If a spool of wire will not be used overnight, it should be returned to its carton and tightly sealed, unless it is in a spool enclosure that provides a dry or protective atmosphere. The 5XXX series electrodes are most likely to develop a hydrated oxide and, when not in use, should be stored in cabinets maintaining a relatively low humidity (less than 35% RH).

1.5 Aluminum Alloy: Principal characteristics and applications

Wrought Alloys

1xxx—Pure Al

- Strain hardenable.
- High formability, corrosion resistance and electrical conductivity.
- Electrical, chemical applications.
- Representative designations: 1100, 1350.
- Typical ultimate tensile strength range: 70–185 MPa.

The 1xxx series represents the commercially pure aluminum, ranging from the baseline 1100 (99.00% min. Al) to relatively purer 1050/1350 (99.50% min. Al) and 1175 (99.75% min. Al). Some, like 1350 which is used especially for electrical applications, have relatively tight controls on those impurities that might lower electrical conductivity. The 1xxx series are strain-hardenable, but would not be used where strength is a prime consideration. Rather the emphasis would be on those applications where extremely high corrosion resistance, formability and/or electrical conductivity are required, e.g., foil and strip for packaging, chemical equipment, tank, car or truck bodies, spun hollowware, and elaborate sheet metal work.

2xxx—Al-Cu Alloys

- Heat treatable.
- High strength, at room and elevated temperatures.
- Aircraft, transportation applications.
- Representative alloys: 2014, 2017, 2024, 2219, 2195.
- Typical ultimate tensile strength range: 185–427 MPa.

The 2xxx series are heat-treatable, and possess good combinations of high strength (especially at elevated temperatures), toughness, and, in specific cases, weldability. They are not resistant to atmospheric corrosion, and so are usually painted or clad in such exposures. The higher strength 2xxx alloys are primarily used for aircraft (2024) and truck body (2014) applications. These are usually used in bolted or riveted construction. Specific members of the series (e.g., 2219 and 2048) are readily welded, and so are used for aerospace applications where that is the preferred joining method.

Alloy 2195 is a new Li-bearing alloy for space applications providing very high modulus of elasticity along with high strength and weldability. There are also high-toughness versions of several of the alloys (e.g., 2124, 2324, 2419), which have tighter control on the impurities that may diminish resistance to unstable fracture, all developed specifically for the aircraft industry. Alloys 2011, 2017, and 2117 are widely used for fasteners and screw-machine stock.

Applications for the 2xxxx series alloys

Aircraft internal structure includes extrusions and plate of 2xxx and 7xxx alloys like 2024, 2124 and 2618. External sheet skin may be aluminum clad 2024 or 2618. The higher purity cladding provides corrosion protection to the Al-Cu alloys that will darken with age otherwise. Heavy dump and tank trucks and trailer trucks employ 2xxx extrusions for their structural members. The fuel tanks and booster rockets of the Space Shuttle are 2xxx alloys, originally 2219 and 2419. Now sometimes Al-Li 2195 alloy is being used for space applications.

3xxx—Al-Mn Alloys

- High formability, corrosion resistance joinability and medium strength.
- Heat transfer, packaging, roofing-siding applications.
- Representative alloys: 3003, 3004, 3005.
- Typical ultimate tensile strength range: 110–283 MPa.

The 3xxx series are strain-hardenable, have excellent corrosion resistance, and are readily welded, brazed and soldered. Alloy 3003 is widely used in cooking utensils and chemical equipment because of its superiority in handling many foods and chemicals, and in builders' hardware. Alloy 3105 is a principal for roofing and siding. Variations of the 3xxx series are used in sheet and tubular form for heat exchangers in vehicles and power plants. Alloy 3004 and its modification 3104 are among the most widely used aluminum alloys because they are drawn and ironed into the bodies of beverage cans.

Typical applications
Automotive radiator heat exchangers are of alloys like 3002. Alloy 3003 is used for tubing in commercial power plant heat exchanger. The bodies of beverage cans are made up of alloys 3004 or 3104.

4xxx—Al-Si Alloys

- Heat treatable.
- Good flow characteristics, medium strength.
- Pistons, complex-shaped forgings.
- Representative alloys: 4032 and filler alloy 4043.
- Typical ultimate tensile strength range: 172–379 MPa.

Of the two most widely used 4xxx alloys, 4032 is a medium high-strength, heat-treatable alloy used principally for forgings in applications such as aircraft pistons.

Alloy 4043 on the other hand is one of the most widely used filler alloys for gas-metal arc (GMA) and gas-tungsten arc (GTA) welding 6xxx alloys for structural and automotive applications. Good flow characteristic provided by the high silicon content, which in the case of forgings ensures the filling of complex dies and in the case of welding ensures complete filling of crevices and grooves in the members to be joined. For the same reason, other variations of the 4xxx alloys are used for the cladding on brazing sheet, the component that flows to complete the bond. The several classes of 4xxx alloys in widest use are represented by the following applications: Refrigerator coolant circulation system in brazed unit of high-Si brazing alloy sheet. Alloy 4043 is one of the most widely used weld filler wires.

5xxx—Al-Mg Alloys

- Strain hardenable.
- Excellent corrosion resistance, toughness, weldability.
- Building and construction, automotive, cryogenic and marine applications.
- Representative alloys: 5052, 5083, 5754.
- Typical ultimate tensile strength range: 124–352 MPa.

Role of Magnesium
Magnesium is the main alloying addition in the 5000 series. Alloy 5052 contains about 2.5% magnesium and is, used for ductwork. The higher you go with Mg, greater is the as-welded strength or annealed strength of the base metal. Up in the 4 to 5.5% magnesium range, there are 5086, 5083 and 5456 aluminum.

These alloys are used in the construction of boat hulls, railroad coal cars, truck bodies and tank trailers. These higher strength alloys are developed following the development of gas metal arc process.

6000 series materials possess good weldability provided adequate filler metal is fed into the joint. A problem can be that fabricators do not open the joint sufficiently to get enough filler metal in there to avoid cracking. Using 4000 series filler metal, it is best to allow about 50% dilution of the filler into the weld metal composition. With a 5000 series filler metal, one needs about 70% filler metal dilution. If a 3.18 mm thick section of 6061 aluminum is welded with insufficient 4043 filler metal, it will result in cracking.

The 2000 series presents a wide range of weldability from one end to the other. The alloys with the lowest solidus temperatures are the most difficult to weld from the standpoint of cracking. Alloy 2219, with its copper content of 6.25%, is the best. Its solidus temperature is 538°C. The corrosion resistance of the 2000 series may not be as good as 6000 series. Other 2000 series alloys have lower solidus temperatures and are more susceptible to weld cracking than 2219. Alloy 2014 is weldable to some degree. But the other 2000 series alloys are considered unweldable by arc welding.

Al-Mg alloys of the 5xxx series are strain hardenable, and have moderately high strength, excellent corrosion resistance even in salt water, and very high toughness even at cryogenic temperatures to near absolute zero. They are readily welded by a variety of techniques. As a result, 5xxx alloys find wide application in building and construction, highways structures including bridges, storage tanks and pressure vessels, cryogenic tankage and systems for temperatures as low as –270°C (near absolute zero), and marine applications. Alloys 5052, 5086, and 5083 are the workhorses from the structural standpoint, with increasingly higher strength associated with the increasingly higher Mg content. Specialty alloys in the group include 5182, the beverage can end alloy, and thus among the largest in tonnage, 5754 for automotive body panel and frame applications.

Care must be taken to avoid use of 5xxx alloys with more than 3% Mg content in applications where they receive continuous exposure to temperatures above 100°C. Such alloys may become sensitized and susceptible to stress corrosion cracking. For this reason, alloys such as 5454 and 5754 are recommended for applications where high temperature exposure is likely.

Examples of applications for the broadly used 5xxx series of alloys

High speed ships employ 5083-H113/H321 machined plate for hulls, hull stiffeners, deck and superstructure. Single or multiple hull high-speed ferries, employ several Al-Mg alloys, 5083, and 5454 as sheet and plate (along with 6xxx extruded shapes, to be described next) with all-welded construction. Alloy 5083 is also used for the huge spherical tanks for shipboard transport of liquefied natural gas. The Foresmo bridge in northern Norway is an excellent example of the use of Al-Mg alloys for built up girders systems. This gives a major advantage of replacement aluminum bridges, the ability to pre-fabricate the spans and move them in place quickly, minimizing the disruption to traffic. Rugged coal cars are provided by welded 5454 alloy plate construction. The demands of high-humidity and water exposure in offshore oilrigs are met with 5454, 5086, and 5083 Al-Mg alloy welded construction. Automotive structures are likely to employ increasing amounts of 5754–0 formed sheet for parts such as internal door stiffeners or the entire body-in-white.

6xxx—Al-Mg-Si Alloys

- Heat-treatable.
- High corrosion resistance, excellent extrudability, moderate strength.
- Building and construction, highway, automotive, marine applications.
- Representative alloys: 6061, 6063, 6111.
- Typical ultimate tensile strength range: 124–400 MPa.

The 6xxx alloys are heat treatable, and have moderately high strength coupled with excellent corrosion resistance. They are readily welded. A unique feature is their extrudability, making them the first choice for architectural and structural members where unusual or particularly strength or stiffness criticality is important. Alloy 6063 is perhaps the most widely used aluminum alloy because of its extrudability. It was a key in the recent all-aluminum bridge structure erected in a few days in Foresmo, Norway, and is the choice for the Audi automotive space frame members. Higher strength 6061 alloy finds broad use in welded structural members such as truck and marine frames, railroad cars, and pipelines.

Among specialty alloys in the series, 6066-T6, with high strength for forgings, 6111 for automotive body panels with high dent resistance, and 6101 and 6201 for high strength electrical bus and electrical conductor wire, respectively.

Most important applications for Al-Mg-Si alloys
The power of extruded Al-Mg-Si alloys is the "put-the-metal-where-you-need-it" flexibility that these alloys and the extrusion process provide. Roof structures for areanas and gymnasiums are usually of 6063 or 6061 extruded tube, covered with 5xxx alloy sheet. Geodesic domes, such as the one made originally to house the "spruce Goose" in Long Beach, California, the largest geodesic dome ever constructed, at 1000 ft across, 400 ft high. An integrally stiffened bridge deck shape usually produced in 6063, used to produce replacement bridge decks, readily put in the roadway in hours. The new magnetically levitated trains in Europe and Japan employ bodies with 6061 and 6063 structural members. Extruded Al-Mg-Si alloys may make up the entire frame of motorcycles or cars e.g. Audi A-8 body. Welded 6063 extrusions combined with 5083 tube make up the front and rear axle bodies for the BMW Model 5.

7xxx—Al-Zn Alloys

- Heat treatable.
- Very high strength, special high toughness versions.
- Aerospace, automotive applications.
- Representative alloys: 7005, 7075, 7475, 7150.
- Typical ultimate tensile strength range: 221–607 MPa.

The situation is much the same with the 7000 series of heat-treatable aluminum alloys. For example, Alloys 7075 and 7079, due to their low solidus temperatures and wide melting ranges, are not considered weldable by the GMA methods. The aluminum-lithium group of alloys is also part of the 7000 as well as that of the 2000 series. Admittedly, the aluminum-lithium alloys are a hot topic nowadays, but they are very proprietary. So far, they have not been used to any great degree.

Both 7005 and 7039 aluminum are mainly zinc-magnesium alloy systems. They both have low copper contents and possess good weldability. These alloys are fabricated into such products as sporting goods, military bridges and automobile bumpers. One rule of thumb is that if the copper is under 0.10%, then the material is weldable. Most of the other 7000 series aluminum alloys, especially those used in aircraft, have significant amounts of copper. There the solidus temperature goes down to about 476°C. They also have a very wide melting range. No wonder alloy 7075 cracks when it is welded. This same alloy, however, is easily welded by resistance spot welding or by flash welding, where compressive loads are applied during welding.

Alloys 7005 and 7004, are basically extrusion alloys. Motor bicycles are being produced from alloy 7005 tubing. Rail cars are also being weld fabricated from the same alloy. In Europe, alloy 7005 extrusions are being used in rapid transit cars and in bus chassis. This alloy is not so widely used in the United States. One reason seems to be, these high zinc-magnesium content alloys are anodic to other aluminum alloys. For example, if 7005 is welded to 6061 aluminum, the 7005 will pit and corrode, in the presence of an electrolyte, as it protects the 6061 material. As a result, the trend in this country is to keep 7005 within its own family, to weld 7005 to 7005.

Alloy 7039 is a basic alloy used for aluminum armour in military tanks and personnel carriers. It is only available in plate form. Alloys 7004, 7005 and 7039 are normally welded with 5356, 5183 or 5556 filler alloys.

The 7xxx alloys are heat treatable and among the Al-Zn-Mg-Cu versions provide the highest strengths of all aluminum alloys. There are several alloys in the series that are produced especially for their high toughness, notably 7150 and 7475, both with controlled impurity level to maximize the combination of strength and fracture toughness. The widest application of the 7xxx alloys has historically been in the aircraft industry, where fracture-critical design concepts have provided the impetus for the high-toughness alloy development. These alloys are not considered weldable by routine commercial processes, and are regularly used in riveted construction. The atmospheric corrosion resistance of the 7xxx alloys is not as high as that of the 5xxx and 6xxx alloys, so in such serivce they are usually coated or, for sheet and plate, used in an aluminum-clad version. The use of special tempers such as the T73-type are required in place of T6-type tempers whenever stress corrosion cracking may be a problem.

Applications of 7xxx alloys
Aircraft structures are of 7xxx alloy sheet or extrusion construction. Alloys like 7075-T73 or high-toughness alloys like 7050 or 7475 are among the principal choices. Lightweight 7029 or 7129 bumper structures help improve our gas mileage.

8xxx—Alloys with Al + Other Elements
(not covered by other series)

- Heat treatable.
- High conductivity, strength, hardness.
- Electrical, aerospace, bearing applications.
- Representative alloys: 8017, 8176, 8081, 8280, 8090.
- Typical ultimate tensile strength range: 117–241 MPa.

The 8xxx series are alloys with lesser used alloying elements such as Fe, Ni and Li. Each of these alloying elements is used for the particular characteristics it provides the alloys. Fe and Ni provide strength with little loss in electrical conductivity. So they are used in a series of alloys represented by 8017 for conductors. Li in alloy 8090 provides exceptionally high strength and modulus. This alloy is used for aerospace applications where increase in stiffness combined with high strength reduces component weight.

Cast Alloys
In comparison with wrought alloys, casting alloys contain larger proportions of alloying elements such as silicon and copper. This results in a largely heterogeneous

cast structure, i.e. one having a substantial volume of second phases. This second phase material warrants careful study, since any coarse, sharp and brittle constituent can create harmful internal notches and nucleate cracks when the component is later put under load. The fatigue properties are very sensitive to large heterogeneties. As will be shown later, good metallurgical and foundry practice can largely prevent such defects. The elongation and strength, especially in fatigue, of most cast products are relatively lower than those of wrought products. This is because current casting practice is as yet unable to reliably prevent casting defects. In recent years however, innovations in casting processes have brought about considerable improvements, which should be taken into account in any new edition of the relevant standards.

2xx.x—Al-Cu Alloys

- Heat treatable/sand and permanent mold castings.
- High strength at room and elevated temperatures; some high toughness alloys.
- Aircraft, automotive applications/engines.
- Representative alloys: 201.0, 203.0.
- Approximate ultimate tensile strength range: 131–448 MPa.

The strongest of the common casting alloys is heat-treated 201.0/AlCu4Ti. Its castability is somewhat limited by a tendency to microporosity and hot tearing, so that it is best suited to investment casting. Its high toughness makes it particularly suitable for highly stressed components in machine tool construction, in electrical engineering (pressurized switchgear casings), and in aircraft construction. Besides the standard aluminum casting alloys, there are special alloys for particular components, for instance, for engine piston heads, integral engine blocks, or bearings. For these applications the chosen alloy needs good wear resistance and a low friction coefficient, as well as adequate strength at elevated service temperatures. A good example is the alloy 203.0/Alcu5NiCo, which to date is the aluminum casting alloy with the highest strength at around 200°C. Landing flap mountings and other aircraft components are made in alloys of the 201.0 or in A356.0 types.

3xx.x—Al-Si + Cu or Mg Alloys

- Heat treatable/sand, permanent mold, and die castings.
- Excellent fluidity/high strength/some high-toughness alloys.
- Automotive and applications/pistons/pump/electrical.
- Representative alloys: 356.0, A356.0, 359.0, A360.0.
- Approximate ultimate tensile strength range: 131–276 MPa.

The 3xx.x series of castings are one of the most widely used because of the flexibility provided by the high silicon contents and its contribution to fluidity plus their response to heat treatment which provides a variety of high-strength options. Further the 3xx.x series may be cast by a variety of techniques ranging from relatively simple sand or die casting to very intricate permanent mold, lost foam/lost wax type castings, and the newer thixocasting and squeeze casting technologies. Among the workhorse alloys are 319.0 and 356.0/A356.0 for sand and permanent mold casting, 360.0, 380.0/A380.0 and 390.0 for die casting, and 357.0/A357.0 for many type of casting including especially the squeeze/forge cast technologies. Alloy 332.0 is also one of the most frequently used aluminum casting alloys because it can be made almost exclusively from recycled scrap. Some of the typical applications are: Inner turbo frame for a Mercedes truck. Gearbox casing for passenger car in alloy pressure die cast 380.0. Rear axle housing of 380.0 sand casting. Automotive components of 357.0 and A357.0 sand and permanent mold castings. 3xxx.x castings made by the lost-foam process, providing exceptional detail and complexity. A356.0 cast wheels are now widely used in industry.

4xx.x—Al-Si Alloys

- Non-heat treatable/sand, permanent mold, and die castings.
- Excellent fluidity/good for intricate castings.
- Typewriter frames/dental equipment/marine/architectural.
- Representative alloys: 413.0, 443.0.
- Approximate ultimate tensile strength range: 117-172 MPa.

Alloy B413.0/AlSi12 is notable for its very good castability and excellent weldability, which are due to its eutectic composition and low melting point of 570°C. It combines moderate strength with high elongation before rupture and good corrosion resistance. The alloy is particularly suitable for intricate, thin walled, leak-proof, fatigue resistant castings.

5xx.x—Al-Mg Alloys

- Non-heat treatable/sand, permanent mold, and die.
- Tougher to cast/provides good finishing characteristics.
- Excellent corrosion resistance/machinability/surface appearance.
- Cooking utensils/food handling/aircraft/highway fittings.
- Representative alloys: 512.0, 514.0, 518.0, 535.0
- Approximate ultimate tensile strength range: 117-172 MPa.

The common feature, which the third group of alloys have, is good resistance to corrosion. Alloys 512.0 and 514.0 have medium strength and good elongation, and are suitable for components exposed to sea water or to other similar corrosive environments. These alloys are often used for door and window fittings, which can be decoratively anodized to give a metallic finish or in a wide range of colors. Their castability is inferior to that of the Al-Si alloys because of its magnesium content and consequently long freezing range. For this reason it tends to be replaced by 355.0/Alsi5Mg, which has long been used for similar applications. For die castings where decorative anodizing is particularly important, the alloy 520.0 is the most suitable.

7xx.x—Al-Zn Alloys

- Heat treatable/sand and permanent mold cast (harder to cast).
- Excellent machinability/appearance.
- Furniture/garden tools/office machines/farm/mining equipment.
- Representative alloys: 705.0, 712.0.
- Approximate ultimate tensile strength range: 207-379 MPa.

Because of the increased difficulty in casting 7xx.x alloys, they tend to be used only where the excellent finishing characteristics and machinability are important.

8xx.x—Al-Sn Alloys

- Heat treatable/sand and permanent mold castings (harder to cast).
- Excellent machinability.
- Bearings and bushings of all types.
- Representative alloys: 850.0. 851.0.
- Approximate ultimate tensile strength range: 103-207 MPa.

Like the 7xx.x alloys, 8xx.x alloys are relatively hard to cast and are used only where their unique machining and bushing characteristics are essential. In concluding this section on casting, it is worth noting that conventional die casting tends to yield parts with relatively low elongation values, which are therefore unsuitable for safety-critical components. In recent years, higher pressure types of casting (e.g., squeeze casting and thixocasting) have been developed to a commercial level. As a result, elongation values of well over 10% are now attainable, together with higher strengths. This considerably widens the range of application of aluminum alloy castings.

2
Surface Preparation Guidelines

Quality of aluminum welding depends upon cleanliness and dryness of the metal with a thin oxide film. The basic cause of porosity in aluminum welds is hydrogen. Any moisture or hydrocarbons in the arc area will dissociate to provide hydrogen. The moisture penetrates and causes inward oxide growth. This hydrated-oxide layer possesses chemically combined water to cause porosity. Also, the thickened oxide is difficult to remove by the arc action; it melts at 2038°C, which is three times the melting point of the aluminum alloy, and prevents proper fusion.

When material is received, it should be properly stored. This will minimize future cleaning operations. Good storage practice is to position the sheets vertically and far enough apart to permit moisture to run off and to allow air circulation to dry the surfaces. Prior to welding, the first operation should be to remove all grease, oil, dirt, paint or other surface contaminants that can generate hydrogen gas or interfere with weld fusion. Degreasing may be done by wiping, spraying, or dipping in a solvent or by steam cleaning.

The naturally formed aluminum oxide film is removed by the arc of the inert gas shielded process or by the fluxes used in by other joining methods. Mechanical oxide removal, although not as consistent as chemical means, is usually satisfactory if performed properly. Wire brushing is by far the most widely used methods. Stainless steel brushes with 0.254 to 0.381 mm, diameter bristles give good scratching action. The brush must be kept clean of all contaminants and it should be used with a light pressure to avoid burnishing the aluminum surface and entrapping oxide particles.

Pre-weld Cleaning

Pre-weld cleaning of aluminum is essential for optimum weld quality. Pre-cleaning requirements are especially stringent prior to direct current electrode negative GTAW, because under such conditions, the arc exerts no cleaning action. Surface contaminants that should be removed from the base metal include dirt, metal particles, oil and grease, paint, moisture, and thick oxide coatings. Another source of contamination is oxide film on the filler metal. The thicker the oxide, the greater its adverse effect on weld-metal flow and solidification and the greater the risk of porosity. Any foreign material that remains on the surface to be welded is a potential source of unsound welds. For best results, all cleaning and oxide removal should be done immediately before welding.

First, the work-metal surface should be cleaned of contaminants. The following manual cleaning methods can be used for small production runs. Dirt can be removed easily by washing and scrubbing with a detergent solution; drying is necessary to ensure that no moisture is present on the surfaces to be welded. Removal of grease and oil can be accomplished by swabbing with cloths soaked in an approved nontoxic solvent.

Next, thick oxide layers should be mechanically removed with a wire brush, steel wool, mill filer, portable milling tool, or a scraper. The use of abrasive paper or grinding disks alone is not recommended, because particles of the abrasive may become embedded in the aluminum and, unless subsequently removed, can cause inclusions in the weld. Wire brush bristles preferably should be 0.254 to 0.381 mm in diameter and made of stainless steel to minimize iron oxide pickup.

Chemical removal of oxides can be accomplished by immersion in solutions of the butyl alcohol-phosphoric acid type. After the chemical treatment, the parts should be washed thoroughly with water and dried with air. For thick and persistent oxide coatings, immersion in a 5% sodium hydroxide solution at 150°F for about 30 seconds is recommended. Degreasing and chemical cleaning should be done before the parts are assembled for welding. Freshly machined and freshly filed surfaces are the cleanest and are often specified when the ultimate in weld quality is demanded.

3

Joint Design and Edge Preparation

In general joint design for aluminum alloys is similar to that for steel. Some recommended joint designs for arc welding of aluminum are shown in Fig. 3.1. When using direct current electrode negative (DCEN straight polarity) GTAW, the root face can be thicker and the grooves narrower. Welding a lap joint or a T-joint on one side only is not recommended. A small continuous filler weld on each side of the joint is preferable.

3.1 Edge Preparation and Assembly

Materials up to about 10 mm thick can be sheared to a reasonably square edge that can be cleaned readily. Dull or improperly designed tools result in lapping of material on prepared edges that can trap lubricant, which can cause weld porosity. The extra time needed to ensure a close fit is often less than the extra time required in welding an improperly prepared assembly. A very close fit of the edges is also essential when GTAW without the addition of filler metal.

3.2 Preheating

Preheating may be advantageous for GTAW with alternating current of parts thicker than about 5 mm and GMAW of parts thicker than about 25 mm. Gas Tungsten Arc welding with DCEP is limited to thin material, and preheating is not necessary with this process. Thick parts also should not be preheated when GTAW using DCEN, because of the high heat input provided to the work. The preheating temperature depends on the job. Often 100°C is sufficient to ensure adequate penetration on weld starts, without readjustment of the current as welding progresses.

3.3 Fixtures

Design of fixtures is based on the expectation that dimensional changes in welding aluminum alloys are twice as great as in the welding of steel. The coefficient of expansion of aluminum is about twice that of steel, and its melting point is about half that of steel. In general, the amount of expansion is inversely proportional to the speed of welding.

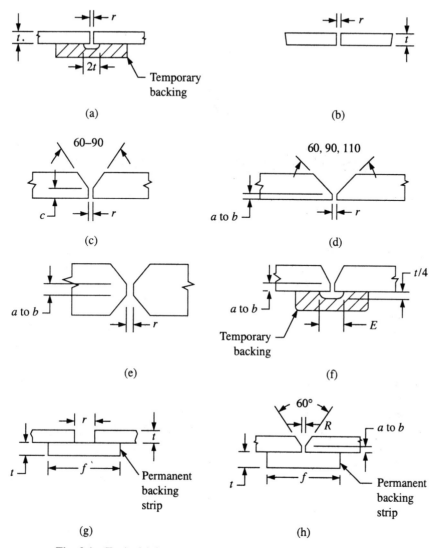

Fig. 3.1   Typical joint geometries for arc welding of aluminum

Although rigid clamping reduces distortion, the inability of the weldment to contract, caused by the restraint, may induce residual stress as high as the yield strength of the base metal and may also result in cracking. Hold-down fingers should be designed with minimum separation between members, and welding should be done in the minimum number of passes.

4

Arc Welding Power Sources

Arc welding power sources can be as varied as the welding processes. To meet the unique electrical requirements of various arc welding processes many types of power sources are necessary. Here the power sources needed for shielded metal arc (SMAW), gas metal arc (GMAW) and gas tungsten arc (GTA) will be described.

The line voltage available is too high to use directly in arc welding. Arc welding involves low-voltage, high current arcs between an electrode and the work piece. Therefore the first function of arc welding power source is to reduce the high line voltage to a suitable voltage range of usually 20 to 80 volts. Either a transformer, solid-state inverter or a motor-generator set can be used to obtain this rated terminal or open circuit voltage appropriate for arc welding. The same power source also provides a high welding current generally ranging from 50 to 1500 amperes. The typical output of a power source may be alternating current (ac) or direct current (dc). The output power may have a characteristics of either constant-current, constant-voltage or both. It may also provide a pulsing output mode. A schematic presentation of the basic elements of an arc welding power source is shown in Fig. 4.1.

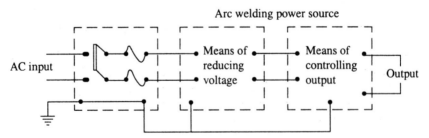

Fig. 4.1 Basic element of an arc welding power source

Apart from the welding processes that use pulsed current, welding power sources are commonly classified as constant current or Drooping and constant potential or Flat. Such classifications are based on the static volt-ampere characteristics of the power supply and not on the dynamic characteristics.

Generally, the word 'constant' is partially true constant-potential power sources are usually much closer to constant voltage output than constant-current sources are to constant-current output. However specialized power sources are available which can hold the outputs truly constant. The fast response solid-state sources can provide power in pulses over a broad range of frequencies. These are known as pulsed power sources.

Volt-Ampere Characteristics

The effectiveness of all welding power sources is determined by two kinds of operating characteristics, each of which affects welding performance differently. They are defined as static and dynamic characteristics. Both affects arc stability but in different ways depending on the welding process. Static output characteristics can be readily measured under steady state conditions by conventional test procedures using resistive loads. A set of output-voltage versus output-current characteristics curves also known as volt-ampere curves are usually used to describe the static characteristics.

The dynamic characteristics of an arc welding power source is determined by measuring the transient variations in the output current and voltage that appear in the arc. These characteristics describe instantaneous variations. Their occurrence is very short of the order of 0.001 second. In general welding arcs operate in changing conditions. In particular, transients occur,

- during the striking of an arc,
- during rapid changes in arc length,
- during the metal transfer across the arc,
- in the case of AC welding, during arc extinction and re-ignition each half-cycle.

These arc transients can occur in 0.001 second, the time interval during which significant change in ionization of arc column occurs. The power source must respond rapidly to these demands. This makes it important to have a dynamic characteristics control of the power source. The static volt-ampere characteristics have little significance in determining dynamic characteristics of an arc welding system. The dynamic characteristics of an arc welding power source are influenced by the following design features

- local transient energy storage, such as dc series inductance or parallel capacitance circuits,
- feedback controls in automatically regulated systems,
- modifications of waveforms or circuit operating frequencies.

Controlling these characteristics results in improvement in arc stability which leads to,

- improvement in uniformity of metal transfer,
- reduction in metal spatter,
- reduced weld pool turbulence.

Manufacturers of welding power supplies give the static volt—ampere characteristics, however there is no universally accepted method by which dynamic characteristics can be specified.

4.1   Constant Current Power Source

The typical volt-ampere characteristics of a conventional constant current power supply is shown in Fig. 4.2. This is also known as drooping power source because of the drooping nature of the volt-ampere curve. These power sources may have open circuit voltage adjustment in addition to output current control. By changing either of these controls, the slope of the ampere curve will change. Constant-current power sources are typically used for SMAW, GTAW and SAW. These machines can be inverters, transformer rectifier or generators.

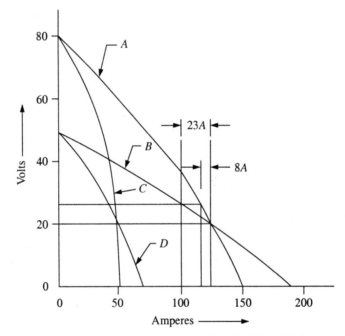

Fig. 4.2   Typical volt-ampere characteristics of a "drooping" power source with adjustable open circuit voltage

The open circuit voltage of constant-current, rectifier type power sources vary depending on the intended welding application. They range from 50 V to 100 V. The effect of the slope of the V-A curve on power output is shown in Fig. 4.2. The curve-A indicates an open circuit voltage of 80 V. An increase in arc voltage from 20 V to 25 V i.e. for a 25% change, it would result in a decrease in current from 123 A to 115 A i.e. 6.5%.

The relative change in current is much smaller. Therefore in case of SMAW, a small change in the arc length due to manual operation resulting in a change in arc voltage will lead to a much smaller change in welding current. The net effect is electrode melting rate remains fairly constant. If the open circuit voltage is set at 50 V, curve-B, for the same change of arc voltage i.e. from 20 to 25 V, the change in current would be from 123 A to 100 A i.e. 19%, which is much higher than the previous case. Hence, a less skilled welder would prefer a power source with much stiffer V-A curve, so that the current remains more or less constant even if there are fluctuations in arc length. However, a skilled welder might prefer a power source with much flatter V-A curve, because he can substantially vary the welding current hence the metal deposition by changing the arc length. This could be useful in case of out-of-position welding because a welder could control the electrode melting rate and molten pool size. Current control, as shown by curves C and D, is used to provide lower power output. It would result in V-A curves with higher slope resulting in a more constant-current output for greater changes in voltages.

4.2 Constant Voltage Power Source

A typical volt-ampere (V-A) curve for a constant-voltage power source is shown in Fig. 4.3. Ideally the V-A curve should have been flat i.e. parallel to current axis. However because of internal electrical impedance there is a minor drop in the output voltage with increasing current. This leads to a V-A curve with slight downward slope. This slope therefore can be changed by changing the internal impedance.

It can be seen from the Fig. 4.3, that with increase or decrease in arc voltage to A or C i.e. ±5 V or 25% change, will produce a much larger change in current (±100 A or 50%). Hence welding processes having constant electrode feeding mechanism, such as gas metal arc (GMAW), submerged arc (SAW) or flux cored arc (FCAW) maintaining constant arc length are suitable for constant—voltage power sources.

A slight change in arc length will cause a substantial change in welding current. This will automatically increase or decrease the electrode melting rate to regain the desired arc length. This effect is known as self regulation. The difference between static and dynamic characteristics of a power supply can be explained from Fig. 4.3. For example during short circuiting metal transfer in GMAW, the welding electrode tip touches the weld pool, causing a short circuit. At this point the arc voltage almost drops to zero and there is a rapid increase in current. However this rapid increase in current gets limited by the circuit inductance.

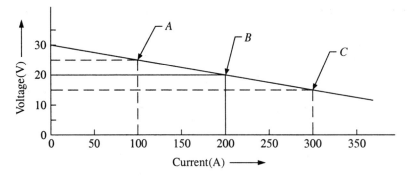

Fig. 4.3 Volt-ampere output relationship for a constant voltage power source

If the power supply responded instantly, very high current would immediately flow through the welding circuit quickly melting the short-circuited electrode tip and freeing it with an explosive force. This would result in dispelling the weld metal as spatter. Dynamic characteristics designed into these power sources controls the rate of current change, thus decreasing the explosive force.

A typical constant-voltage (constant-potential or flat) power source has a negative slope of 1–2 V per 100 A. Machines with *V-A* curves having slopes upto 8 V per 100 A are still referred to as constant-voltage power supplies.

4.3   Pulsed Mode Power Source

Pulsed power supply is used to reduce the arc power while retaining the desirable spray transfer. The concept is based on the fact that metal transfer from the electrode wire takes place in either of the two following ways:

- Spray transfer. In this mode when the current exceeds a critical level, metal transfer takes place in the form of a few hundred drops per second.
- Globular transfer. In this mode when the welding current is below the critical level, the metal transfer takes place in the form of less than 10 drops per second.

This critical current at which transition from globular to spray transfer takes place, is known as transition current. By pulsing the current between these two regions, the desirable quality of spray transfer can be achieved while reducing the average current significantly. This also reduces deposition rate. Because of this GMAW with pulsed power supply can be used to weld in all positions as well as thinner plates even sheet metal. The current level during the globular transfer interval is kept sufficiently low so as to avoid any metal transfer however high enough to sustain ionization in the arc region. This current is more commonly known as 'background current' which helps in keeping the arc alive. Whereas in the spray interval, the current is raised above the transition current for a sufficient time, enough to allow transfer of one or two droplets. This current is known as 'pulse current'. Power supplies are designed with necessary controls to deliver the controlled output for pulsed GMAW.

With solid-state power sources such as inverters, it is now possible to have a control over all the pulse variables, peak current, background current, peak current time and background current time (Fig. 4.4). By controlling these variables it is possible to have a total control of metal transfer, allowing only single drops to transfer per pulse, while retaining the background current.

One of the major drawbacks of a pulsed power source is, it is difficult to set the variables for a given welding requirement. Once the variables are set, it is also difficult to reset them should any of the variables needs to be changed. For example, a number of variables need to be reset if the wire feed speed is changed. Such problems can be overcome by having electronic and microprocessor controls to set the optimum pulse conditions for a given wire feed speed setting. Fig. 4.5 is a diagrammatic representation of the circuit elements of a typical synergic pulsed power source. The word 'synergic' means 'several things acting as one'.

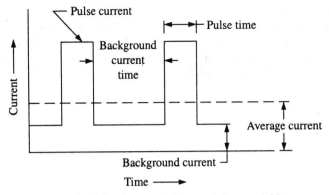

Fig. 4.4 Pulsed gas metal arc welding variables

In synergic pulsed GMAW machines, the pulse variables are set automatically based on the required wire feed speed for a given joint geometry. Through the use of electronic controls, it is possible to choose a variety of synergic curves to

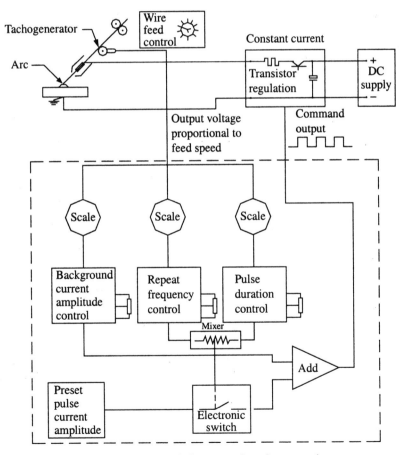

Fig. 4.5 Basic circuit for synergic pulse operation

satisfy particular applications. Synergic power supplies can be so designed that it can transfer single drops with each pulse by making instantaneous adjustments of the pulse frequency and width depending on the voltage sensed across the arc. Pulsed GMAW power sources typically range up to 500 A peak current and the frequency various from 60 to 200 Hz.

5

Welding Parameters

Weld quality and weld deposition rate both are influenced very much by the various welding parameters and joint geometry. Essentially a welded joint can be produced by various combinations of welding parameters as well as joint geometries. These parameters are the process variables which control the weld deposition rate and weld quality. The variables are:

- welding current,
- arc voltage,
- welding speed,
- electrode feed speed,
- electrode extension (length of stickout),
- electrode diameter,
- joint geometry.

Each of these parameters has influence to a varying degree on the deposition rate, weld bead shape, depth of penetration, cooling rate and weld induced distortion. Hence to achieve a sound welded joint with adequate metal deposition rate and minimum weld induced distortion, a proper understanding of the effects of these parameters or process variables is needed.

The general effect of these parameters will be dealt with individually in this chapter. In the subsequent chapters their effects for various welding methods will be studied in detail.

5.1 Welding Current

For any given electrode the melting rate is directly related to the electrical energy supplied to the arc as well as the polarity of the electrode in case of DC power supply. Part of this energy is used to melt a portion of the base metal, part goes to melt the electrode (q_e) and rest gets dissipated by way of conduction (q_{cp} and q_{ce}), convection (q_v) and radiation (q_r) (Fig. 5.1). This energy is directly proportional to the welding current as can be seen in the following:

$$Q = I. V \qquad \text{J/s}$$

or
$$Q = I^2 \cdot R_a \qquad \text{J/s}$$

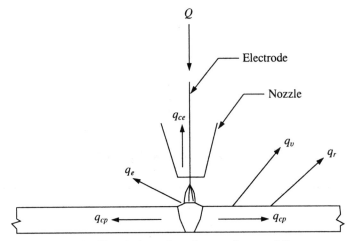

Fig. 5.1   Heat balance in submerged arc welding

where Q = electrical energy consumed
I = welding current
V = arc voltage
R_a = arc resistance

Welding current is the most influential variable in a welding process. It controls the electrode melting rate and hence the deposition rate, the depth of penetration, and the amount of base metal melted. If the current is to high at a given welding speed, the depth of fusion or penetration will be too great. For thinner plates, it tends to melt through the metal being joined. It also leads to excessive melting of electrode resulting in excessive reinforcement. This means additional heat input to the plates being welded leading to increased weld induced distortions. On the other hand, if the welding current is too low, it may result in lack of fusion or inadequate penetration. Welding current can be either DC or AC, both have their respective advantages and disadvantages.

Direct current always provides a steadier arc and smoother metal transfer as compared to AC arc. The DC arc produces good wetting action by the molten weld metal and uniform weld bead size even at low amperage. For this reason DC is particularly suited to thin section welding. Direct current produces a better quality in vertical and overhead welding. The disadvantage of DC welding is arc blow when magnetic metals are welded. The advantage of alternating current over direct current are two fold. One is the absence of arc blow and the other is the cost of power source. However while selecting the type of welding current for a quality job, all operating factors need to be considered.

5.2   Welding Arc Voltage

Arc voltage is the voltage between the electrode and the job during welding. It is determined by arc length for any given electrode. Open circuit voltage, on the other hand, is the voltage generated by the power source when no welding is done (Fig. 5.2).

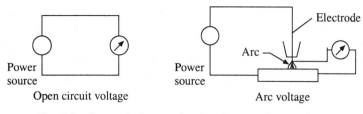

Fig. 5.2   Concept of open circuit voltage and arc voltage

Open circuit voltage generally vary between 50 v to 100 v, whereas arc voltages are between 17 v to 40 v. When the arc is struck, the open circuit voltage drops to the arc voltage and the welding load comes on the power supply.

This arc voltage depends on the arc length and the type of electrode. If the arc length is increased there will be an increase in the arc resistance resulting in higher drop in voltage, or in other words the arc voltage will increase and the current will decrease. This drop in current will depend on the slope of the volt-ampere curve as shown in Fig. 4.3.

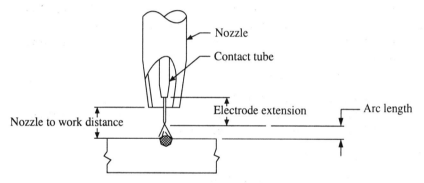

Fig. 5.3   Gas metal arc welding terminology

Arc length

The arc length is the distance from the molten tip of the electrode core wire to the surface of the molten weld pool (Fig. 5.3). Proper arc length is important in obtaining a sound welded joint.

Metal transfer from the electrode tip is not a smooth, uniform action. There is a variation in instantaneous arc voltage as a metal droplet is transferred through the arc. However such variations will be minimal when welding is done with proper amperage and arc length. Constant and consistent electrode feed is required to maintain a constant arc length.

The correct arc length is a function of electrode diameter, covering composition, amperage and welding position. With increase in electrode diameter and amperage the arc length should also increase. As a general rule the arc length should not be more than the electrode diameter.

A short arc tends to become erratic and may lead to short circuit during metal transfer. On the other hand a long arc will lack direction and intensity. This results in scattering of molten metal as metal transfer takes place from electrode to the weld pool. Thus heavy spatter may occur with a loss in deposition rate. Arc voltage and arc length are related to each other. With all variables held constant, arc voltage is directly related to arc length. Though the arc length as a welding parameter is to be controlled to obtain a quality welding, it is much easier to monitor arc voltage. Because of this arc voltage as a welding parameter is more commonly used.

The weld bead appearance depends on the arc voltage. An increase in arc voltage tends to flatten the weld bead, and increase the width of the fusion zone. Excessive high voltage may cause porosity, spatter and may lead to undercut. Reduction in arc voltage may cause narrower weld bead with a higher crown and deeper penetration. Hence trials are essential to obtain an optimum arc voltage.

5.3 Welding Speed

Welding speed is the linear rate at which the arc moves along the weld joint. Welding speed is particularly important because it controls the actual welding time and hence it has a direct effect on the cost. Therefore the speed generally is fixed in mechanized welding while the other parameters like current and/or arc voltage are varied to control the weld deposit and weld quality.

The effect of welding speed for any given combination of welding current and arc voltage generally conforms to a general pattern.
If welding speed is increased,

- heat input per unit length of welded joint decreases,
- less filler metal is deposited, resulting in less weld reinforcement,
- undercut, arc blow, porosity and uneven bead shape may result.

If welding speed is decreased,

- filler metal deposition rate increases,
- heat input rate increases,
- weld bead gets wider and more convex,
- penetration decreases with further decrease in speed,
- a large molten pool resulting in a rough bead and possible slag inclusion.

With all other welding parameters held constant, weld penetration attains a maximum at an intermediate speed of welding. For excessive slow welding speed when the arc strikes a rather larger molten pool, the penetrating force of the arc gets cushioned by the molten pool of metal. With excessive welding speed, there is a substantial drop in thermal energy per unit length of welded joint, results in undercutting along the edges of the weld bead because of insufficient deposition of filler metal to fill the path melted by the arc. Within limits the welding speed can be adjusted to control weld size and depth of penetration.

5.4 Electrode Extension

The electrode extension, also known as length of stickout, is the distance between the end of the contact tube and the end of the electrode as shown in Fig. 5.3. An increase in the length of stickout will result in an increase of electrical resistance.

This leads to resistance heating of the electrode between the contact tube and the arc. This gives rise to an additional heat generation and increases the electrode melting rate. Hence, the longer the extension, the greater is the heat generation and voltage across the electrode thus reducing the power delivered to the arc. This reduces the arc voltage which in turn reduces the bead width as well as weld penetration (Fig. 5.4).

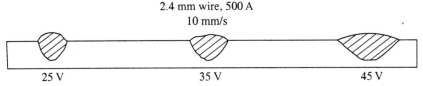

2.4 mm wire, 500 A
10 mm/s

25 V   35 V   45 V

Fig. 5.4   Effect of Arc voltage variations on weld bead shape and penetration (steel welding)

Lower arc voltage also will lead to an increased convexity of the weld bead. Hence to maintain the proper bead shape and desired penetration with an increased length of stickout for higher melting rate, the voltage setting on the machine should be increased to maintain proper arc length. However, at current densities over 125 A/mm^2, the length-of-stickout becomes an significant variable. An increase of 25% to 50% in deposition rate can be achieved by using long electrode extensions without increasing the welding current.

This increase in deposition rate is always accompanied by decrease in penetration. Therefore, when deep penetration is desired long electrode extension is not recommended. On the other hand in case of thinner material, when there is a possibility of melt-through, a longer electrode extension will be beneficial. However, as the electrode extension increases, it becomes more and more difficult to maintain the correct position of the electrode tip with respect to the joint to be welded.

5.5   Electrode Diameter

The electrode diameter influences the weld bead configuration. It has a direct effect on weld penetration and deposition rate (Fig. 5.5).

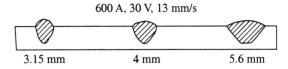

600 A, 30 V, 13 mm/s

3.15 mm   4 mm   5.6 mm

Fig. 5.5   Effect of electrode size on weld bead shape and penetration (steel welding)

At any given current, a smaller diameter electrode will have a higher current density causing a higher deposition rate compared to an electrode of large diameter. Therefore a larger diameter electrode requires a higher minimum current than a thinner electrode to achieve the same metal transfer characteristics. Hence a higher diameter electrode will produce higher deposition rate at higher current.

If a desired electrode feed rate is higher (or lower) than the feeder motor can deliver, changing to a larger (or smaller) size electrode will permit the desired deposition rate. In case of poor fit-up, a larger diameter electrode is better than smaller ones for bridging large root openings.

Gas Metal Arc Welding
By far most used process for welding aluminum is gas metal arc welding (GMAW). It combines speed with versatility and the ability to make high-quality welds. The GMAW process is applicable to heavier thicknesses of aluminum. It is much faster than gas tungsten arc welding. Several factors should be mentioned with respect to GMAW aluminum. The electrode wire must be clean. If porosity occurs, it is possible that it came from moisture absorbed in the oxide coating of the electrode wire. Pure argon is normally used for GMAW of aluminum. On occasion, leaks in the gas system, in the gun or cable assembly, will allow air to be drawn into the argon which will cause porosity. Gas purge control and post-gas flow should be used. The angle of the gun or torch is critical. A 30 degree leading travel angle is recommended. The electrode wire tip should be oversized for aluminum. The wire feeding equipment for aluminum welding must be in good adjustment for efficient wire feeding. Nylon liners should be used in cable assemblies. Proper drive rolls should be selected for the aluminum wire and for the size of the electrode wire. The spool gun is used for the small-diameter electrode wires. Water-cooled guns are required except for low-current welding.

Both the constant current power source with matching voltage-sensing wire feeder and the constant-voltage power source with constant speed wire feeder are used for welding aluminum. The CV system is preferred when welding on thin material and using small diameter electrode wire. It provides better arc starting and regulation. The CC system is preferred when welding thick material using larger electrode wires. The CC power source with a moderate droop of 15 to 20 V per 100 A and with a constant speed wire feeder provides the most stable power input and provides the highest weld quality.

Gas Metal Arc Welding of aluminum has the same advantages over earlier welding processes, as does gas tungsten arc welding (GTAW), i.e. no need for flux, little spatter, ability to weld in all positions and an arc that removes the oxide film to allow coalescence of joint edges and filler. It has some additional advantages in that much higher welding speeds, greater depths of penetration and narrower heat-affected zones are possible with GMAW than with GTAW. The GMAW arc is a very efficient and concentrated heat source in comparison with other welding arcs.

The GMAW process is limited to using direct current, electrode positive (DECP) power. With DCEP, the arc removes the aluminum oxide film, most of the arc heat is generated at the electrode. This concentrates the welding heat for higher thermal efficiency and narrower heat-affected zones. Metal transfer is produced by a magnetic force generated by the welding current, which pinches off droplets of molten metal and accelerates them to high speeds across the arc length. The inertia of this metal as it reaches the weld pool gives the GMAW arc its characteristic deep penetrating ability and the positive metal transfer under

magnetic force overcomes gravity and makes it possible to weld in all positions. GMA welding requires that the electrode wire feed rate balance the melting rate in the arc. If more filler is supplied than is being melted, the arc gets progressively shorter and finally stubs out on the base metal surface. If less filler is supplied than is melted, the arc gets progressively longer and finally reaches the point of causing a meltback. In either case, the welding operation shuts down. Two approaches are to synchronize wire feed and melting rate. One is to vary the wire speed to balance the current and the other is to vary the current to suit the wire speed. Commonly used to vary the current to compensate for wire speed fluctuations.

Thicknesses
Thicknesses of aluminum alloys commonly joined by GMAW range from 3 mm up to the maximum plate thickness available (several mm). In this thickness range, GMAW is capable of high quality weld deposits.

Welding Speeds
Welding speeds up to 24 mm/s are obtained with semiautomatic welding, and speeds for machine and automatic welding can be as high as 75 mm/s. Maximum welding speeds commensurate with the application are always desirable when welding aluminum alloys. Rapid cooling after welding, which results from high welding speeds, produces fine-grain weld deposits and retards the formation of low-melting constituents at the grain boundaries.

6

Power Sources

GMAW of aluminum requires a suitable and stable arc, one that adequately cleans the base metal in and around the weld pool, deposits a uniform weld bead without undercutting, gives uniform penetration, does not disrupt the inert gas shielding envelope and is easily and consistently initiated. The most important factors in this respect are electrode wire speed uniformity, joint geometry, weld pool dynamics and arc initiation. *Only direct current electrode positive (DCEP), which gives good penetration and a cathodic cleaning action at the work surface, is used in GMAW of aluminum alloys.* The principal advantage of the constant current (CC) power source is that the current is kept constant even when electrode speed fluctuations or other causes vary the voltage somewhat. The constant current results in uniform penetration. Its disadvantages are that the wire speed must not vary and it is a little more difficult to initiate the arc than with a constant voltage machine. The principal advantage of the constant voltage (CV) power source is that the arc length is kept constant even when the electrode wire speed varies. This tends to prevent meltbacks. It is also easier to initiate the arc. Its main disadvantage is that any variation in electrode wire speed cause a change in welding current and thus in penetration. These responses are not necessarily visible to the welder and the arc may be quite unstable without he or she being aware of it. The standard for steel welding has been the CV machine, which minimizes the risk of meltbacks and simplifies arc striking. In the 1980s, inverter power source emerged. This new power source type rectifies to DC first, converts to a high-frequency AC (in the kilohertz range) for voltage transformation and then rectifies to provide direct current for welding. These power sources give a "stiff" arc that makes it easier to penetrate to the root of joints, especially into deep grooves and other difficult-to-weld joints. These machines are small, often light enough to be portable and tend to be less costly than conventional power sources.

6.1 Arc Stability and Voltage/Current Characteristics

A stable arc is necessary to get good arc cleaning and good shielding gas coverage of the arc and weld pool to prevent porous welds. Short arcs can cause excessive reinforcement. Wire feed speed uniformity is necessary for arc stability. However, aluminum electrode wire is not as easy to feed at constant rate as is steel due to its physical properties. Feeding problems tend to be more especially with push feeders and long flexible conduits. Even if the electrode enters the conduit at a uniform speed, it may exit the gun in an erratic or pulsing fashion because there is enough clearance between wire and conduit liner for intermittent buckling. However momentary they may be, the arc will respond to any speed changes and the result will be arc instability. A CV power source compensates by varying the current, which varies arc penetration, but the arc remains stable. A CC power source compensates by varying the arc length, which is immediately perceived as arc instability, and leads to a meltback. It thus appears that *for aluminum, inverter power sources may have the best features of CC and CV machines without their disadvantages.*

6.2 Joint Geometry and Voltage/Current Characteristics

Joint geometry is a factor in the selection of the most suitable power source. Because penetration is primarily a function of welding current and the CC power source maintains a uniform current, it is not significantly affected by joint geometry. On the other hand, a CV power source may have difficulty in maintaining a stable arc with some joint configurations. In a narrow deep groove, it may have difficulty deciding whether the arc length is to be registered from the bottom of the groove or from one or the other side. This produces an arc that hunts violently in terms of actual physical arc length and the resultant weld is uneven both in penetration and in bead formation. Another drawback with a CV power source is its inability to weld over tack welds while continuing to provide uniform penetration. As soon as a tack is met, it immediately disturbs the arc, which reduces its penetrating power. Thus, it tends to over lay the tack weld without penetrating it adequately. A CC machine is not subject to this fault because its arc length is determined by wire speed, which is not affected by the presence of a tack weld.

The recommended joint geometry for arc welding aluminum is similar to those for steel. *Aluminum joints have smaller root openings and larger groove angles than are generally used for steel since aluminum is more fluid and welding gun nozzles are larger.* Typical joint geometry for arc welding aluminum are shown in Fig. 3.1. V-groove joint designs are adequate for butt joints that are accessible from both sides. As a rule, *a 60-degree groove angle is the minimum practical size for section thickness greater than 3.0 mm.* Thick sections may require even larger groove angles, such as 75 or 90 degrees, depending upon the welding process.

6.3 Pool Dynamics and Voltage/Current Characteristics

The welding arc determines its length from the end of the electrode wire to the weld pool surface. Unfortunately, the weld pool is not always flat, level and quiescent. In the molten form, aluminum is very fluid and disturbances during welding can easily upset the pool surface. A high current GMAW arc exerts significant arc pressure on the pool causing it to be depressed under the arc. If the base metal does not melt at a constant rate, the arc pressure can change, causing the pool surface to react. This, in turn, will change the arc length and the welding system will attempt to correct for the change. To do this, a CV power source will make large current swings that will result in large arc pressure changes that tend to perpetuate the problem. With a CC machine, such disturbances seldom occur but, in any case, tend to damp out quickly.

One extreme form of pool instability causes a defect called "cabbage leafing" or puckering. It is so sever that the metal in the pool rolls under the arc and in so doing disturbs the gas shielding and allows air into it. The result is metal deposit that has the appearance of a plowed field and cannot be called a weld. Its onset can usually be attributed to either the arc speed being too high for the welding conditions or the gas shield having been disturbed. Once initiated, this condition can usually only be arrested by stopping the welding operation and before restarting, correcting the conditions causing it.

7

Electrode Wires for Gas Metal Arc Welding

Electrode-wire feed should be selected so that the wire is consumed as fast as it emerges from the welding torch without extending more than 10 mm beyond the shielding gas nozzle. The torch is tilted not more than 10° forehand. The arc length that should be used is governed by the metal thickness, the type of filler-metal alloy and the welding current. When making small fillet welds and welding narrow-groove butt joints, a short arc is preferred. Arc length is usually 3 mm to 9 mm. The wire sizes chosen for each application depend on the requirements and welding conditions for that application.

The electrode wire is a critical component in the GMAW process. Since it acts as a filler metal, it must not add hydrogen to the weld pool. The surface cleanliness of GMAW wire is very important if weld porosity is to be avoided. Hence it calls for correct protection of the electrode wire during use. As soon as it is removed from its carton and plastic enclosure, it is in danger of becoming contaminated by the shop atmosphere. It is recommended to return the spool to its package when no welding is being done for more than a few hours.

8

Electrode Feeder and Gun

Most GMAW machines have a feeder and gun designed to work together. The combination has a number of necessary functions:

- To drive electrode wire from a spool through the gun so it arrives at the arc at a uniform and controlled speed.
- To transfer welding current to the electrode wire as close as possible to the arc.
- To guide the electrode wire in the arc so that the welder can, by positioning the gun, be certain where the weld bead will form.
- To deliver shielding gas to the arc area so as to blanket the arc and the weld pool.
- To provide cooling for the power cable and the gun.
- To provide circuitry so the welder can initiate and interrupt welding power, electrode wire feed, shielding gas flow and cooling water flow as desired. Electrode wire feed there are three main systems, push feeders, pull feeders and push-pull feeders. For aluminum, push feeders are limited to a practical length of flexible conduit of about 3 m because of the flexibility of the aluminum electrode and its tendency to buckle. Pull feeders are also limited to about the same length. On the other hand, push-pull feeders can have conduits up to 7.5 m long if the system is designed to maintain light tension on the electrode during its passage through the conduit.

  A push feeding unit contains a spool of electrode wire, a drive motor and drive rolls and controls for welding power, wire feed, gas flow and water flow. Feed rolls are usually mounted in pairs with a powered grooved roll and a flat pressure roll. This push-type feeder arrangement results in two advantages over conventional drive rolls, i.e., it straightens the electrode wire and imparts a vibration to it, both of which make it slide more easily through the flexible conduit.

8.1 Guns

The nozzle must be large enough to cover the arc zone with shielding gas without being so large that gas is wasted or that the welder has difficulty seeing the arc and weld pool. The recommended nozzle size is usually included in the welding procedure.

8.2 Contact Tubes

It picks up several hundred amperes through rubbing contact at two or three points. Some tubes are only about 25 mm long while others are up to 100 mm. There seems to be an advantage to the longer tubes in giving a more stable arc. Welders prefer a gun that does not require special tools but gives easy access to the tube for its removal.

9

Shielding Gas

The shielding gases used for GMA welding are argon, helium or a mixture of the two. Not only does the shielding gas protect the arc and the weld pool from the surrounding air, it also affects the characteristics of the arc resulting from the ionizing potential of the particular gas.

GMAW arcs in argon give narrow fusion at the limit of penetration while those in helium give a much wider fusion shape at this location. By adjusting the composition of the shielding gas (which is either argon, helium or a mixture of the two), some control is provided over the distribution of heat to the weld. This, in turn, influences the shape of the weld-metal cross section and the speed of welding. Adjustment of the cross section by modification of the shielding-gas composition can control penetration without changing electrode melting rate to facilitate evolution of gas from the weld metal. However, use of helium only as a shielding gas usually results in an unstable arc and is seldom used for that reason.

Typical cross sections of bead welds that were made using argon, helium, argon-helium mixtures and CO_2 are shown in Fig. 9.1. The relatively narrow and deep argon-shielded weld may favour gas entrapment. The deep and narrow projection at the center of the argon-shielded weld can be used to gain penetration at a given level of current density. However, when making a joint with one weld on each side, the weld placement must be carefully controlled to ensure overlap of the two passes at the center. For a specific arc length, the addition of helium will increase the arc volts by 2 or 3 volts as compared to

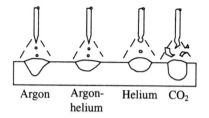

Argon   Argon-   Helium CO_2
       helium

Fig. 9.1   Bead contour and penetration patterns for various shielding gases

argon alone. Since the weld profile broadens and deepens to a maximum at 75% He-25% Ar, this mixture has been a common choice for achieving sound welds when welding heavy plate from both sides. Argon is used for most semiautomatic and some automatic welding because it gives adequate penetration, clean welds and is cheaper than helium. The purity of the shielding gas is of utmost importance. Only gases having a dew point of 60°C or better should be used and care must be taken to prevent contamination. Dust, dirt and moisture can accumulate in the cylinder fitting, which should be carefully cleaned and blown out before use. Plastic hose is recommended.

Shielding Gas Composition

The primary function of shielding gas is to protect the arc and the molten weld pool from atmospheric oxygen and nitrogen. If not properly protected it forms oxides and nitrites and results in weld deficiencies such as porosity, slag inclusion and weld embrittlement. In addition to this, the shielding gas performs a number of important functions,

- forms the arc plasma,
- etabilises the arc roots on the material surface,
- ensures smooth transfer of molten metal droplets from the wire to the weld pool.

Thus the shielding gas and its flow rate have a substantial effect on the following:

- Arc characteristics
- Mode of metal transfer
- Penetration and weld bead profile
- Speed of welding
- Cleaning action
- Weld metal mechanical properties

The principal gases used in GMAW are shown in Table 9.1. The gases used for short circuiting transfer are shown in Table 9.2. General purpose shielding gases for GMAW are mixtures of argon, oxygen and CO_2 and special gas mixtures may contain helium. Gases, which are normally used for various materials, are:

Steels	Non-ferrous metals
CO_2	Argon
Argon + 2 to 5% O_2	Helium
Argon + 5 to 25% CO_2	

9.1 Argon

Argon is generally preferred when welding thinner metal, mainly because of its lower arc heat. In addition, argon results in a smoother and more stable arc than helium, and thus much less weld spatter is obtained.

Argon-Oxygen
It is difficult to strike an arc in pure argon atmosphere [901]. The addition of oxygen to the argon shielding gas generally improves the ability to strike an arc and provides a more stable metal transfer. The addition of oxygen to argon

Table 9.1   GMAW shielding gases for spray transfer

Metal	Shielding gas	Thickness	Advantages
Aluminum	100% Argon	0 to 25 mm	Best metal transfer and arc stability; least spatter.
	35% Argon, 65% Helium	25 to 76 mm	Higher heat input than straight argon; improved fusion characteristics with 5XXX series Al-Mg alloys.
	25% Argon, 75% Helium	Over 76 mm	Highest heat input; minimises porosity
Magnesium	100% Argon	–	Excellent cleaning action.
Carbon Steel	95% Argon +3.5% Oxygen	–	Improves arc stability; products a more fluid and controllable weld puddle; good coalescence and bead contour; minimises undercutting; permits higher speeds than pure argon.
	90% Argon + 8%/10% CO_2	–	High-speed mechanised welding; low-cost manual welding.
Low-alloy Steel	98% Argon –2% Oxygen	–	Minimises undercutting: provides good toughness.
Stainless Steel	99% Argon –1% Oxygen	–	Improves arc stability; produces a more fluid and controllable weld puddle, good coalescence and bead contour; minimises undercutting on heaviour SS.
	98% Argon –2% Oxygen	–	Provides better arc stability; coalescence, and welding speed than 1% oxygen mixture for thinner stainless steel materials.
Nickle, Copper, and their alloys	100% Argon	up to 3.2 mm	Provides good wetting; decreases fluidity of weld metal.
	Argon-Helium	–	Higher heat inputs of 50 and 75 percent helium mixtures offset high heat dissipation of heavier gases.
Titanium	100% Argon	–	Good arc stability; minimum weld contamination; inert gas backing is required to prevent air contamination on back of weld area.

Table 9.2   GMAW shielding gases for short circuiting transfer

Metal	Shielding gas	Thickness	Advantages
Carbon steel	75% Argon + 25% CO_2	Less than 3.2 mm	High welding speeds without burn-through; minimum distortion and spatter.
	75% Argon + 25% CO_2	More than 3.2 mm	Minimum spatter; clean weld appearance; good puddle control in vertical and overhead positions.
	Argon with 5–10% CO_2	–	Deeper penetration; faster welding speeds.
Stainless steel	90% Helium-7.5% Argon + 2.5% CO_2	–	No effect on corrosion resistance; small heat-affected zone; no undercutting; minimum distortion.
Low-alloy steel	60–70% Helium + 25 – 35% Argon + 4.5% CO_2	–	Minimum reactivity; excellent toughness; excellent arc stability, wetting characteristics, and bead contour; little spatter.
	75% Argon + 25% CO_2	–	Fair toughness; excellent arc stability, wetting characteristics, and bead contour; little spatter.
Aluminum, Magnesium, Nickel, Copper and their alloys	Argon & Argon + Helium	Over 3.2 mm	Argon satisfactory on sheet metal; argon-helium preferred on base material.

shielding gas has only a small influence on the arc characteristics [903]. The observed effects of the addition of oxygen to an argon-shielded GMAW process are mainly due to altered transport phenomena at the anode (electrode) and the cathode (work piece weld pool). P.G. Jönsson et. al in [903] has observed that the addition of oxygen leads to improved arc stability by:

- Forming nonmetallic inclusions that initiate arcing,
- Forming oxides that decrease the movement of arc (cathode spots).

Oxygen also affects the geometry of the weldment by

- Forming films that improve the wetting conditions of the weld pool,
- Affecting the surface tension gradient such that deeper and more narrow weld pools are formed.

These effects of oxygen addition to argon are summarized in Fig. 9.2.

The addition of oxygen does not affect the electrode melting rate significantly [904]. Frseman and Eagar in [905] found no difference in droplet transfer frequency for the pure argon gun and the 98% argon and 2% oxygen mixture. However, they noticed that the average drop size and neck diameter were smaller and were more elongated in case of pure argon as a shielding gas.

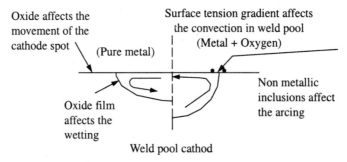

Fig. 9.2   **The influence of oxygen on the molten pool**

The addition of up to 5% oxygen to an argon shielding gas affects the characteristics of the arc column to only a small extent. The mass flow, momentum flux, electric potential and temperature change by no more that 5% when O_2 is added to argon gas [903]. The influence of O_2 addition on transport phenomena at the anode needs to be further investigated.

Argon-Carbondioxide
Jönsson in [906] found that the droplet diameter for a 95% argon-5% carbon dioxide shielding gas mixture is smaller and the frequency of droplet transfer is higher as compared to that of 98% argon-2% oxygen gas mixture at all current levels. The 95% argon-5% CO_2 gas mixture contains a higher oxygen content than the 98% argon 2% oxygen mixture. As a result of chemical reaction between oxygen and the liquid metal droplet, the surface tension of liquid metal decreases with an increase in oxygen concentration [907]. Therefore, 'O_2 richer' 95% argon-5% CO_2 gas mixture lowers the surface tension more than the 98% argon 2% O_2 mixture. The static force balance theory [904, 908, 909] explains the effect of surface tension on droplet detachment. This states that a drop detaches from the electrode tip when the static detaching forces (gravitational, electromagnetic and plasma drag forces) exceed the static retaining force due to surface tension. This is why argon-carbon dioxide gas mixture yields smaller droplet sizes and higher droplet transfer frequencies compared to the argon-oxygen mixture.

Argon-Oxygen-Carbon Dioxide
Shielding gas mixtures of argon with up to 20% CO_2 and 3 to 5% O_2 are versatile. They provide adequate shielding and desirable arc characteristics for spray, short circuiting and pulse mode welding.

Argon-Helium-Carbon Dioxide
This gas mixture, is suitable for short circuiting and pulse mode arc welding of carbon, low alloy and stainless steels. Argon rich mixtures are used for pulse arc welding whereas helium rich mixtures are used for short circuiting arc welding.

9.2   Helium

Helium, because of its greater arc heat, is capable of producing deep penetration desirable in weld deposits in thicker metal. The bead profile with helium shielding is wider and less convex than with argon shielding, and the penetration pattern has a broader underbead. Helium is lighter than argon, requires higher flow rates, and is more expensive. Therefore, helium is seldom used alone.

9.3 Flow Rates

Typical shielding gas flow rates for GMAW of aluminum and aluminum alloys using 1.6 mm diameter electrode wire are:

Shielding gas	Flow rate (ft^3/h)
100% argon	30–70

It provides excellent arc stability, bead shape and penetration. This gas may be used in all welding positions. Helium-argon mixtures are preferred for semiautomatic welding with 5XXX aluminum alloy electrodes and are sometimes used instead of helium to take advantage of the arc stability provided by argon for mechanized and automatic welding.

REFERENCES

901 Doan, G.E. and Myer, J.L., "Arc discharge not obtained in pure argon gas", *Review*, 1932, 40(4); 36–39.

902 Ries, D.E., "Gas metal arc welding of titanium", *MS Dissertation*, 1983, Massachusetts Institutes of Technology, Cambridge, Mass.

903 Jönsson, P.G., Murphy A.B. and Szekely J., "The Influence of Oxygen Additions on Argon-Shielded Gas Metal Arc Welding Processes", *Welding Journal, Research supplement*, Feb. 1995, pp 48–5 to 58.5.

904 Kim, Y.S., "Metal transfer in gas metal arc welding", *Ph.D. Dissertation*, 1989, Massachusetts Institute of Technology, Cambridge, Mass.

905 Freeman, J.J. and Eagar, T.W., "An investigation of gas metal arc welding of steel in pure argon", *Interim Report*, 1992, Materials processing Center, Massachusetts Institute of Technology, Cambridge, Mass.

906 Jonshon, P.G., "Arc properties and metal transfer in gas metal arc welding", *D.Sc. Dissertation*, 1993, Massachusetts Institute of Technology, Cambridge, Mass.

907 Heiple, C.R. and Roper, J.R., "Mechanism for minor element effect on GTA fusion zone geometry", *Welding Journal*, 1982, Vol. 61(4), pp 97-s to 102-s.

908 Greene, W.J. "An analysis of transfer in gas-shielded welding arcs", *Trans. AIEE part 2 7*, 1960, pp 194–203.

909 Waszink, J.H. and Graat, L.H.J., "Experimental investigation of the forces acting on a drop of weld metal", *Welding Journal*, 1983, 62, pp 108-s to 116-s.

10

Arc Characteristics in Gas Metal Arc Welding

Increasing the welding current changes the arc from one producing short circuiting transfer to one producing globular transfer and then to one producing spray transfer. *Spray transfer produced by either a constant-current arc or by a pulsed current arc is used for almost all GMAW of aluminum alloys.* To obtain spray transfer from constant current arcs requires extremely high current densities when welding aluminum alloys. Current densities ranging from 70 to 470 A/mm^2 of electrode cross section have been used. In contrast, current densities for GTAW of aluminum alloys and for GMAW of steel are about 15 A/mm^2. The constant current and current density ranges in which the transition from globular to spray transfer takes place depend on the electrode size and the arc voltage used. For a 1.2 mm diameter electrode and 22 to 31 V, the change in type of metal transfer occurs at about 120 A or about 105 A/mm^2.

Increasing the electrode diameter to 2.4 mm increases the transition current to about 220 A, but decreases the current density to about 46 A/mm^2. Thus when the electrode diameters are larger than standard, the current density for the transfer transition further reduces.

Arc-Length Control

The self-regulation of the arc length can be achieved by two combinations of electrode feed units and power supplies. Most commonly a constant-potential power source in conjunction with a constant-speed electrode feed unit is used to achieve the desired self-regulation. The another option is a Constant-current power source providing drooping volt-ampere curve with a electrode feed unit which is arc-voltage controlled.

With the constant potential power souce along with a constant speed wire feeder unit, any change in the torch position will effect a change in the welding current that exactly matches the change in the electrode stick-out and thus the arc length remains constant for example, by moving away the welding torch from the job will produce an increased stick-out which will reduce the current output from the supply, thereby maintaining the same resistance heating of the electrode.

In the other system, the arc voltage fluctuations due to uneven movement of the torch, readjust the control circuits of the feed unit, which appropriately changes the wire feed speed, and the self-regulation is achieved.

However, in case of aluminum welding, a combination of a constant-current power source and a constant-speed electrode feed unit is used. This combination provides a small degree of self-regulation and requires a high degree of operators skill. However, because of the high thermal conductivity of aluminum a highly skilled welder can have a wide range of control over the current which in turn controls the arc energy.

Metal Transfer Characteristics

The manner, in which the weld metal transfer takes place from the electrode to the weld pool, largely determines the operating features of the process. The principal metal transfer modes are,

- Short circuiting transfer
- Globular transfer
- Spray transfer
- Pulsed transfer

Number of factors influence the type of metal transfer in a GMAW process. The most influential of them are the following:

- Welding current, AC or DC and its magnitude
- Electrode wire diameter
- Electrode composition
- Shielding gas

10.1   Short Circuiting Transfer

Short circuiting transfer mode is used for low current operation with lower electrode diameters. In short circuiting transfer also called 'dip' transfer, the molten metal forming on the tip of the electrode wire is transferred by the wire dipping into the molten weld pool thus causing a momentary short circuit. Metal is therefore transferred only during a period when the electrode tip is in contact with the weld pool and no metal is transferred across the arc gap. This type of metal transfer mode produces a small, fast-freezing weld pool that is generally suited for joining thin sections, for out-of-position welding and for bridging large root openings. Fig. 10.1 indicates the sequence of events during a short circuiting metal transfer and the corresponding current and voltages. The frequency of short circuiting i.e. electrode dipping in molten pool varies from 20 to over 200 times per second. As the electrode wire touches the molten pool, short circuit takes place causing a sharp drop in arc voltage and rise in the welding current (positions A, B, C, D) (Fig. 10.1). The molten droplet of the wire tip gets detached at D and E, initiating an arc as shown in E and F. The rate of current increase should be high enough to heat the electrode and promote metal transfer, at the same time the rate of current increase should be low enough so as to minimize spatter caused by violent separation of the molten droplet from the electrode. The voltage and inductance setting in relation to the wire feed speed is essential to minimize spatter. The surge in current which occurs when the wire dips into the weldpool is, controlled by adjustment of the inducting power source. The optimum inductance setting depends on both the electrical resistance of the welding circuit as well as the melting temperature of the electrode. As the arc is reestablished at E and as the electrode wire feed continues, the tip of the wire starts melting and gets short circuited at H and again the cycle of current increase and metal detachment starts. The open circuit voltage is kept low enough so as to ensure that the drop of molten metal at the wire tip does not detach until it touches the molten pool in the base metal.

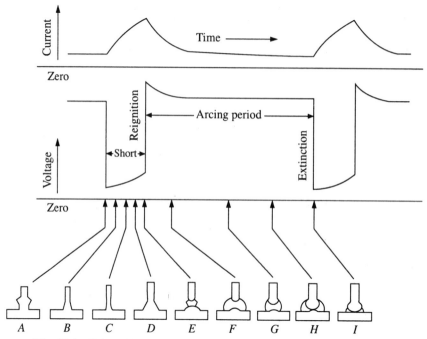

Fig. 10.1   Schematic representation of short circuiting metal transfer

10.2   Globular Transfer

Globular transfer is characterized by a drop size with a diameter greater than that of the electrode. The large drop is easily acted on by gravity, generally limiting successful transfer to the flat position. If vertical welding is done with globular transfer, some of the molten-metal drops will never get to the weld. They get deflected by gravity and becomes weld spatter. Globular transfer takes place with a positive electrode (DCRP/DCEP) when the current is relatively low regardless of the type of shielding gas. At current levels only slightly higher than those used in short-circuiting transfer, globular axially-directed transfer can be achieved in a substantially inert gas shield. At low arc voltages i.e. with too short arc lengths, the molten droplet at the electrode tip grows in size and may even touch the workpiece causing a considerable spatter. The arc length should therefore be long enough or in other words arc voltage should be high enough to ensure detachment of the droplet before it touches the weld pool on the job. However on welding with higher voltage may give rise to lack of fusion, insufficient penetration and excessive reinforcement thus rendering the weld unacceptable.

Transition Point

For a given electrode diameter and an inert gas, the welding current directly controls the size of the droplets and the number of them that are detached from the electrode per unit time as shown schematically in the Fig. 10.2. The molten drop grows in size with increasing current from its lowest value where the arc can be barely sustained. The diameter of the droplet can become several times that of the electrode diameter before they are detached. The transfer rate at low current is only several drops per second. This increase in amperage can go upto a limiting value known as Transition Point. Increasing the current beyond this transition point, the size of the detached drops abruptly decreases (Fig. 10.2) and becomes equal to or even less than the electrode diameter. Simultaneously the rate of detachment of the droplets suddenly increases to several hundred droplets per second. This transition point is associated with a specific current level, for various sizes of electrode diameter. With further increase of current, rate of detachment of droplets increases but the rate of increase is less as compared to that of near transition point.

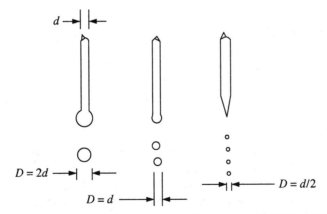

Fig. 10.2 Metal transfer in the spray mode of the pulsed GMAW welding process

The transition current is dependent on the following factors:

- Liquid metal surface tension.
- Inversely proportional to the electrode diameter.
- Length of stickout or electrode extension.
- Electrode wire (filler metal) melting temperature.
- Shielding gas composition.

The metal transfer below this transition point is called globular transfer and the transfer mode with a current above this transition point results in spray transfer. In globular transfer, the droplets detach when their weight exceeds the surface tension of the molten metal that holds the drop on the electrode tip. The electromagnetic force that acts in a direction to detach the drop is small compared to the gravitational force in the current range of globular transfer mode. Typical transition currents for some common metals are shown in Table 10.1. *The type of arc that produces metal transfer by a large drop of molten metal is seldom used when welding aluminum alloys, because the transfer is erratic.*

Table 10.1   Typical transition currents for some common metals

Wire electrode type	Wire electrode diameter (mm)	Shielding gas	Minimum spray arc current (A)
Mild Steel	0.8	98% argon—2% oxygen	150
Mild Steel	0.9	98% argon—2% oxygen	165
Mild Steel	1.1	98% argon—2% oxygen	220
Mild Steel	1.6	98% argon—2% oxygen	275
Stainless Steel	0.9	98% argon—2% oxygen	170
Stainless Steel	1.1	98% argon—2% oxygen	225
Stainless Steel	1.6	98% argon—2% oxygen	285
Aluminum	0.8	Argon	95
Aluminum	1.1	Argon	135
Aluminum	1.6	Argon	180
Deoxidized Copper	0.9	Argon	180
Deoxidized Copper	1.1	Argon	210
Deoxidized Copper	1.6	Argon	310
Silicon Bronze	0.9	Argon	165
Silicon Bronze	1.1	Argon	205
Silicon Bronze	1.6	Argon	270

10.3   Spray Transfer

Spray transfer occurs at welding currents above the transition point as explained before. Above the transition point, the metal transfer occurs in the form of very small droplets that are formed and detached from the electrode tip at the rate of several hundreds per second. These metal droplets are accelerated axially across the arc gap. Spray transfer is associated primarily with the use of inert gases. Either pure argon or argon-rich with 0.5 to 5% oxygen shield gas is used. With such gas mixtures a true spatter-free axial spray transfer becomes possible with DCRP (DCEP) power supply and with current above transition point. The spray transfer yields a highly directed stream of metal droplets with substantial energy so as to have velocities which overcome the effects of gravity. Because of this, spray transfer mode can be used for welding in any position. The metal droplets being very small short-circuit does not occur and hence spatter is virtually eliminated. The spray transfer welding is widely used and almost any metal or alloy can be welded because of the inert gas shielding. However, in this process there is a limiting thickness below which spray transfer can not be used unless the arc is pulsed. Spray transfer welding being a high current, high heat input process, it produces a deep penetration and therefore welding thin plates with this mode may become difficult. The resultant arc forces can cut through thin plates instead of welding them. It also produces high weld metal fluidity with a large weld pool too large to be supported by surface tension in vertical or overhead position, because of very high temperature of the molten weld metal. This high fluidity again makes it difficult to weld in overhead position. This disadvantage is overcome by using pulsed spray welding process.

The spray transfer mode is the one normally used for GMAW of aluminum. The notable characteristics of the spray transfer arc are its stiffness and its narrowness. These advantages are described below.

Arc Stiffness for Deep Penetration

There is no lack of weld penetration when using spray transfer. Even in the low range of welding currents, the use of high current density and small-diameter electrode wire establishes a stable arc column with a well-defined pattern on the workpiece. To ensure fusion at the root of a butt joint, root reinforcement is required, usually 0.8 to 4.8 mm depending on metal thickness and joint design.

Arc Stiffness for Out-of-Position Welding

When using spray transfer, the transfer follows the direction in which the electrode wire is pointed, which makes this type of transfer suitable for out-of-position welding.

Arc Narrowness for Small Fillet Welds

The spray arc has a narrow stable core that concentrates the heat. This property enables fully fused small fillet welds to be made in relatively thick material.

Arc Narrowness for Square-Groove and Narrow-Groove Butt Joints (High Current Density Welding)

The concentrated heat of the spray arc can also be used to weld butt joints with square or narrow grooves, thus reducing the amount and cost of the electrode wire required to make the joint.

At these high current densities, the characteristic hissing noise of the arc is replaced by a crackling noise. These techniques are especially suitable when making square-groove butt joints in base metal from 6 mm to 16 mm thick. Welding is often accomplished in two passes, one from each side, at a much greater speed than is possible at lower current densities. Back gouging is rarely required, and welding in one pass instead of several stringer-bead passes greatly reduces the total heat input. The reduction in heat input results in less distortion and produces better as-welded properties. *Most welds made with the square-butt and high-current techniques are of good quality.*

Thicknesses greater than 16 mm can also be welded with a square-groove butt joint, but the amount of reinforcement may be excessive. To reduce the reinforcement V-grooves can be machined as shown in Fig. 10.3, where two 44 mm plates of 5083 alloy were joined using only one pass from each side, with 450 A welding current, 28V, 2.85 m³/hr of argon for shielding and 1.6 mm diameter ER 5356 electrode wire. The amount of spatter during welding may be more severe with electrodes that contain low-vapor-pressure elements. The aluminum-magnesium alloy electrodes (ER5XXX series) commonly cause the most spatter.

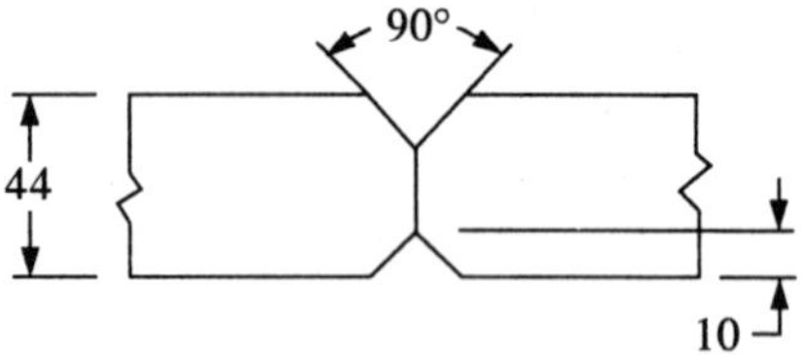

Fig. 10.3 Edge preparation for high current density welding of aluminum plate

The vapor pressure of the magnesium tends to cause disintegration of the droplets as they separate from the electrode tip. This produces small spatter balls that are often thrown clear of the arc. When the arc voltage (arc length) is decreased to below a certain value for a specific electrode and amperage, the size of the droplets will increase and the form of the arc will change. When the arc voltage is in the spray transfer range and the welding current is decreased to below some threshold value, metal transfer will change from spray to globular type. This latter type is not suitable for aluminum welding because fusion with the base metal will be incomplete.

10.4 Pulsed Arc Transfer

The concept of pulsing the current in GMAW welding arose because at the lower end of the current range for each electrode size there is not enough electromagnetic force to pinch off the molten droplet and accelerate it into the weld pool. Below this natural current barrier, transfer takes place only under the effect of gravity or by capillary action if the electrode touches the weld pool. Pulsed arc transfer is a type of spray transfer that occurs in pulses at regularly spaced intervals. In the time interval between pulses, the welding current is reduced and no metal transfer occurs. *The low average current and low heat input associated with pulsed arc welding have allowed the advantages of spray transfer to be extended to the welding of sections thinner than can be spray transfer welded using conventional constant-current power supplies.*

Pulsed arc transfer offers other advantages. Using larger diameter electrode wires, which cost less per kg, are easier to feed, have less current transfer problems in the contact tube, and have a lower probability of weld porosity from surface contamination on the wire because of the lower surface-to-volume ratio of the larger wire. Well-formed root beads are easily made on thin aluminum with pulsed arc welding. Use of the short circuiting transfer has been largely replaced by pulsed arc transfer. The pulsed transfer mode was developed as a means of stabilizing the arc at low current levels within spray transfer range to avoid short-circuiting and spatter. Also the welding difficulties of thin plates and vertical and overhead welding are greatly overcome by using pulsed transfer welding. In pulsed transfer welding the welding current switches automatically from a low level to a higher level in a periodic manner. The power source is adjusted, such that, the lower level current also known as background current is set below the transition point in the globular transfer current range and the higher level is set well above the transition point in spray transfer range. Thus it derives the advantages of both the modes. In this mode spray type metal transfer is achieved by applying pulses of higher level current, each pulse having sufficient force to detach a droplet. The power supplies are specially designed to produce carefully controlled wave forms and frequencies that "pulse" the welding current.

As shown in Fig. 10.4 these power sources provide two levels of current:

- A constant background current which sustains the arc without providing enough energy to cause drops to from on the electrode wire tip,
- A superimposed pulsing current higher than the transition current necessary for spray transfer.

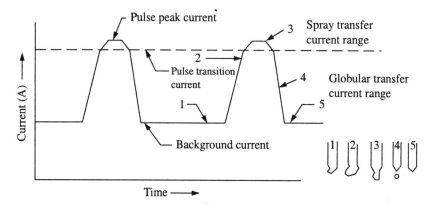

Fig. 10.4   Pulsed-spray arc welding current characteristics

During these pulses, metal transfer takes place. The frequency and amplitude of the pulses control the energy level of the arc and therefore the rate at which the wire melts. Hence with the reduced average arc energy and the wire melting rate thin sections can be effectively welded with the advantages of spray transfer and because of reduced heat input, weld induced distortions are also minimized. Also thick metals in all positions can be welded by this process. Larger diameter electrode wires can be used that cost less than fine wires which also has feeding problems. This mode of transfer calls for a special type of power source. Many variations of these power sources are available. The simplest one provides a single frequency of pulsing (50 or 100 pps) with independent control of the background and pulsing current levels. More sophisticated ones automatically provide the optimum combination of background and pulse for any given wire feed speed. These types of power sources are also known as synergic pulsed power source.

Operating Variables

The weld bead geometry, depth of penetration and overall weld quality depends on the following operating variables:

- Welding current (wire feed speed)
- Polarity
- Arc voltage
- Welding speed
- Length of stickout
- Electrode orientation
- Electrode diameter
- Shielding gas composition
- Gas flow rate.

Welding Current

In normal GMAW, current is controlled by changing electrode feed speed and with all other variables held constant, the welding amperage various nonlinearly with wire feed speed. This relationship of current to wire feed speed for carbon steel, aluminum and stainless steel, electrodes are shown in Fig. 10.5. At low current levels, for different electrode size, there is fairly linear relationship between current and wire feed speed particularly with smaller diameter electrodes. However this gradually becomes non-linear with increase in amperage. This is attributed to resistance heating of the electrode extension beyond the contact tube.

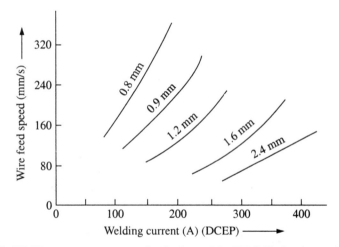

Fig. 10.5   Welding currents versus wire feed speed for ER4043 aluminum electrodes

The wire feed speed S_{wf} can be expressed as a function of welding current I, weld bead reinforcement G, electrode diameter d, heat input per unit length of weld deposit H and open circuit voltage V_0 as given below [1001],

$$S_{wf} = \frac{V_0 GI + nGI^2}{\pi d^2 H/4} \eta \qquad (10.1)$$

where η = effective power supply slope
   G = cross sectional area of deposited metal

The welding current in a GMAW process can be represented by the following equation [1002],

$$I = K_0' + K_1'(L_{ct}) + K_2'V_0 + K_3'S_{wf} \tag{10.2}$$

where K_0' to K_3' are constants.

The current I is valid for any value of reinforcement and heat input which give G/H ratio in the range of 0 to 50 mm^3/J. With all other variables held constant, increase in welding current (electrode feed speed) will result in the following:

- An increase in the depth and width of weld penetration
- An increase in the deposition rate
- An increase in the size of weld bead.

Arc, force and deposition rate being exponentially dependent on current, operation above the transition current may make the arc unstable in the vertical and overhead positions. By reducing the average current with pulsing, the arc forces and deposition rates can both be reduced, allowing welds to be made in all positions and also in thin sections. With metal cored wire additional advantage with stiff power source is that it produces an arc which is less sensitive to changes in electrode stickout and voltage compared to solid wires. Thus the process becomes more tolerant to operator's torch fluctuations.

Polarity

Majority of the GMAW applications use direct current electrode positive (DCEP or DCRP). This gives a stable arc, smooth metal transfer, good weld bead, less spatter and higher depth of penetration for a wide range of welding currents. Whereas in case of direct current electrode negative (DCEN or DCSP) only globular transfer is possible. That is why DCEP is almost always used in GMAW. The electrode melting rate is much higher with DCEN polarity, but the metal transfer mode being globular, this polarity is seldom used. However because of this high deposition rate it has some use in surfacing applications. AC is not used in GMAW because of the tendency of the arc to extinguish as the current passes through the zero point. Some attempts have been made by giving special treatments to the wire surface but it proved to be uneconomical.

Arc Length (Arc Voltage)

The arc length is a critical variable in GMAW. It must be carefully controlled. The arc length determines the current and arc pressure distribution on the weld pool, which in turn determines the size and the shape of the weld fusion zone. When the length is too short, the electrode can contact or short circuit to the weld pool, which will result in a reduced base metal melting, a high and narrow weld deposit, variations in the heat input and increased chances of forming weld defects. On the other-hand excessive arc length causes a wider shallow deposit, makes the arc to wander and increases spatter, and may cause porosity from air aspirated into the shielding gas. The welding arc has a point-to-plane relationship and is thus bell-like in shape with the point at the end of the electrode and the wide portion at the plate surface. Increasing arc length makes the weld bead wider due to the widened arc area at the weld surface and consequently reduces the reinforcement height because the same volume of the filler metal is involved. Conversely, reducing the arc length makes the bead narrower and increases the height of reinforcement. Arc length is an independent variable. Arc voltage depends on the arc length as well as many other variables, such as the electrode composition and dimensions, the shielding gas, welding technique and even the length of the welding cable. The voltage across the contact tube and the work piece is a measurable quality and is generally referred to as arc voltage. (But in reality it is not so.) This voltage can be expressed as

$$V_c = V_t + V_e + V_a \qquad (10.3)$$

Because of these components of V_c, a measure of the true arc voltage alone is difficult. Therefore controlling the total voltage V_c does not ensure a constant arc length. Arc radiation can also be used to control arc length [1003, 1004]. The fraction of power Q_r leaving the arc by radiation can be modeled as

$$Q_r = \eta_r \, IV_a \qquad (10.4)$$

where η_r is fraction of arc power lost by radiation.

As per Watkins et al. [1005], the maximum Q_r is 15% of the total power input to the process. Though the above equation was developed for GTAW, which does not include the complication of metal transfer across the arc, it can also be applied to GMAW to obtain a measure of the power vitiated by the arc. The radiative power in terms of the arc light intensity can be measured by a photodiode. The welding current along with the arc light intensity can be used to measure arc length for a specific shielding gas and electrode composition. The power in the arc column increases with increase in welding current for a given arc length as shown in Fig. 10.6. Also, Fig. 10.7 shows that power increases with increase in

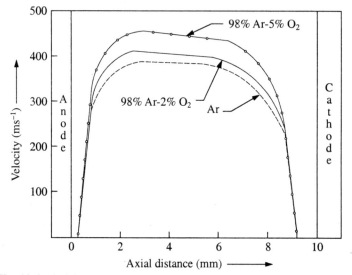

Fig. 10.6   Axial velocity at the centre of the arc as a function of the axial distance from the anode

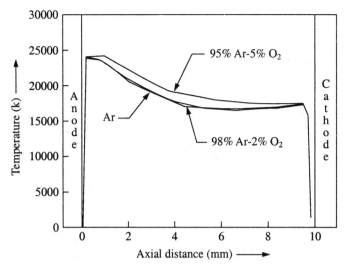

Fig. 10.7 Temperature at the centre of the arc as a function of the axial distance from the anode

arc length at a constant current. The power input on the other hand directly affects the melting of the electrode as well as the formation and detachment of the droplets from the electrode as shown in Fig. 10.8. As a result of which, the arc power input also affects the size and frequency of the droplets, and the solidification of the weld pool, along with the resulting microstructure and mechanical properties of the weld. Since the arc power varies with both arc length and current, it is critical to control not only current but also arc length (arc voltage) specially in case of constant current applications in order to sustain a constant input of power to the system.

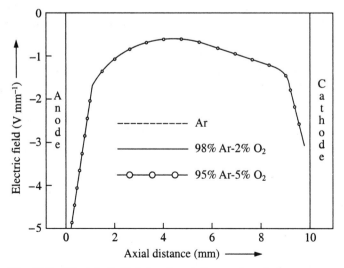

Fig. 10.8 Electric field intensity at the centre of the arc as a function of the axial distance from the anode

Welding Speed

The speed of movement of the welding torch determines the welding speed. Therefore a faster movement of the torch means higher welding speed. As this speed is increased, initially the thermal energy per unit length of weld transmitted to the base metal increases. This happens because the arc at every instant acts directly on the base metal and not on the molten pool. However with further increase in welding speed, less thermal energy gets transmitted to the base metal per unit length of deposited weld metal. Therefore, melting of base metal first increases and then decreases with increase in weld speed. If the speed is further increased, this will cause undercutting along the edges because of insufficient deposition of filler metal being deposited along the welding path. On the other hand when the welding speed is slow, the filler metal deposition per unit length increases with increase in rate of heat input. On further slowing down, the arc impinges on the molten weld pool, rather than on the base metal, thereby reducing the effective penetration and results in a wider weld bead and high heat input. This also leads to higher levels of HAZ, weld induced deformation and residual stresses. Hence, with all other welding parameters held constant, maximum weld penetration is achieved at an intermediate welding speed.

Length of Stickout

Length of stickout is the length of electrode extension from the end of the contact tube. As can be seen from equation (10.3), a voltage drop occurs in the electrode stickout. Hence higher length of stickout will offer higher electrical resistance and will cause a higher voltage drop. This higher resistance will also cause a resistance heating of the electrode resulting in a small increase in electrode melting rate. However, the increase in the length of stickout increases the overall resistance in the welding power circuit, thereby reduces the welding current. The power source generally used in GMAW systems has a flat or CP characteristics and the electrode feed rate is constant. This constant potential system compensates for the variation in the length of stickout by automatically supplying the increased or decreased welding current to attain equilibrium. The power source provides the proper current so that the melting rate is equal to the electrode feed rate. It has been observed that the temperature of the molten droplets is lower in case of the long electrode extension than that of the short extension. This indicates that the high temperature of the electrode extension promotes the detachment of droplets from the electrode end. The increased length of stickout consequently brings about a re-established equilibrium state with a long length of stickout. Fig. 10.9 and Table 10.2, represent the effects of varying electrode extension on the weld bead shape and depth of penetration.

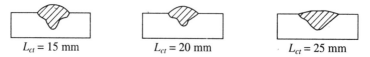

L_{ct} = 15 mm L_{ct} = 20 mm L_{ct} = 25 mm

**Fig. 10.9   Weld geometry according to change of contact
tube-to-workpiece distance L_{ct}**

Table 10.2 Measured weld size and welding currents according to various contact tip-to-workpiece distance L_{ct}

L_{ct} (mm)	Bead width (mm)	Bead height (mm)	Joint penetration (mm)	Welding current (A)
15	11.32	3.12	5.05	300
20	12.60	2.82	3.95	280
25	13.20	2.79	2.84	260

The welding current determines the welding power keeping welding voltage constant. The largest bead width is obtained with longest electrode extension. Whereas depth of penetration distinctly increases with decrease in length of stickout. This occurs due to the constriction of the distributed heat and current flux by the decreased arc length, which results in a larger electromagnetic force, which promotes the convection effect in the weld pool. Thus the length of stickout is an important operating variable which can control the formation of welds.

Electrode Orientation
The electrode orientation with respect to the weld joint affects the weld bead shape as well as weld penetration. The influence of electrode orientation on weld bead shape and its penetration is more than that of arc voltage or welding speed. The electrode orientation is defined in two ways,

- By the relationship of the electrode axis with respect to the direction of travel. This is known as the travel angle.
- The angle between the electrode axis and the adjacent work surface. This is called work angle.

When the electrode points against the direction of travel, the technique is called backhand welding with drag angle. When the electrode points in the direction of travel, it is called forehand welding with a lead angle. Fig. 10.10 indicates the

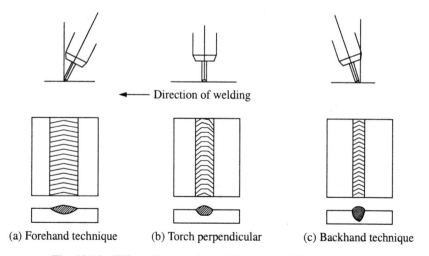

(a) Forehand technique   (b) Torch perpendicular   (c) Backhand technique

Fig. 10.10 Effect of electrode position and welding technique

electrode orientations and their effect on weld profile and penetration. Backhand welding with a drag angle of about 25° in the flat position of the job, gives the maximum penetration. When it is changed to forehand welding with a lead angle, with all other conditions unchanged, the penetration decreases and the weld bead becomes wider and flatter. The drag technique apart from deeper penetration also produces a more convex, narrower bead, a more stable arc, and less spatter on the work-piece. For all positions, the electrode orientation is generally backhand with a drag angle in the range of 5 to 15° for achieving good control and shielding of the molten weld pool. However, for aluminum, a lead technique is preferred. The lead technique provides a cleaning action ahead of the molten weld metal, which promotes better wetting and reduces base metal oxidation. For fillet welds in horizontal position, the electrode should be positioned about 45° to the vertical member, the work angle to be 45° as shown in Fig. 10.11.

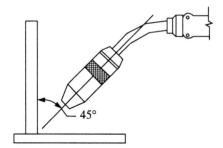

Fig. 10.11   Normal work angle for fillet welds

Electrode Diameter
The weld bead shape also depends on the size of the electrode wire i.e. the electrode diameter. A large diameter electrode requires higher minimum current as compared to an electrode of smaller diameter for the same metal transfer characteristics. Higher currents yield additional electrode melting and more fluid and larger weld deposited. At the same time for larger diameter electrodes the required higher current also results in higher deposition rates as well as greater penetration. However overhead and vertical welding are usually done with smaller diameter electrode with lower currents. The transition currents for aluminum vary with wire diameter. The relationship between the wire diameter and the transition current I_C below which spray transfer ceases to exist for a 22v arc is given by

$$I_C = 68\,d + 30 \qquad\qquad (10.5)$$

where d = electrode diameter in mm.

As the welding current should be more than the transition current to have a spray transfer, welding of thin materials requiring low current should be done with small diameter wires.

REFERENCES

1001 Smart H.B., and Einerson C.J. "A model for heat and mass input control in GMAW", *Welding Journal Research Supplement,* May, 1993, pp. 2175–2295.

1002 SAS Institute, Inc. SAS user's Guide Statistics, 1982 Edition, Carry, N.C.

1003 Johnson, C.A. and Sciaky, A.M. 1966, "System for controlling arc length of welding arcs" U.S. Patent 3236997.

1004 Madigan, R.B. and Quin, T.P., "Sensing droplet detachment and electrode extension for control of GMAW" *Conf. Proc. 3rd ASM Int. Trends in Welding Research,* 1992, pp. 999–1002, ASM International Materials Park, Ohio.

1005 Watkins, A.D., et al., *2nd International Conference on Trends in Welding Research,* 1989, pp. 19–23, ASM International, Materials Park, Ohio, 206.

11

Weld Backing for Gas Metal Arc Welding

Backing bars are commonly used for gas metal arc welds in butt joints, as this permits welding to be accomplished at higher speed, with less operator skill and less control of welding conditions, especially when using spray transfer for joining thin sections. Steel is the material most often used for temporary backing when welding aluminum alloys. Carbon steel is often used but stainless steel is used when lower thermal conductivity is required in the backing or to avoid iron oxide contamination of the weld. Austenitic stainless stee. backing bars have reasonable life against arc damage and being non-magnetic, do not produce arc blow. Their use minimizes the possibility of iron or rust pickup in the root bead.

Aluminum backing with a hard anodic coating provides adequate chilling; an added advantage is that the arc cannot strike the aluminum backing and cause damage, because the anodic coating is an excellent dielectric. Backing bars may be temporary, permanent, or integral. Temporary backing should be grooved to allow the root surface of the weld to protrude beyond the plane of the back surface of the workpiece, thereby ensuring adequate penetration. This groove should be shallow (0.25 to 0.76 mm) and wider than the width of the root surface of the weld. Too wide a groove provides insufficient support for the metal under the hold-down clamps.

12

Welding Procedures

Current, voltage (arc length) and speed are the principal parameters of GMA welding. Welding current determines arc heat, which fuses and penetrates the metal. Voltage or the arc length determines the arc force.

A short arc focuses the available force on a small area of the weld pool thus giving deeper penetration and narrower welds. Table 12.1 illustrates the effect of arc length on the weld. Welding speed is also a contributing factor in determining penetration and weld shape.

Table 12.1   Effect of arc length on weld

Characteristic	With short arc	With long arc
Penetration	Deep	Shallow
Weld width	Narrow	Wide
Reinforcement	Higher	Flatter
Pool surface	Depressed	Flatter
Spatter	More	Less
Arc noise	Crackling	Humming
Weld porosity	More	Less

With a CC machine, adjusting the power source sets the current. The arc length is set by adjusting the wire feed speed. Whereas with a CV machine, the current is set by adjusting the wire feed speed and arc length by adjusting the output voltage of the power source. Constant-voltage power and constant-speed electrode drives are normally used with small diameter electrodes less than 1.2 mm. The electrode feed rate is adjusted to obtain the desired welding current for good fusion and penetration. Arc voltage is adjusted to give a spray transfer mode of filler metal. A constant-current power source and constant-speed electrode drive should be used with 1.2 mm and larger diameter electrodes. The welding current is set at the desired value, and the arc voltage is set by adjusting the wire feed speed to the desired arc length. The arc length (voltage) is critical with respect to good fusion with the groove faces. If the voltage is too low, short circuiting will take place between the electrode and the weld pool. Typical GMAW parameters using small diameter electrodes are given in Tables 12.2 and 12.3.

Table 12.2 Typical welding parameters for GMAW of aluminum alloys in argon shielding

Section thickness (mm)	Welding position	Joint geometry	Root opening (mm)	No. of weld passes	Electrode diameter (mm)	Welding current (DCEP), (A)	Arc voltage (V)	Shielding gas flow rate (L/min)	Travel Speed (mm/sec)
1.6	F	A	0	1	0.8	70–110	15–20	12	10.5–19.0
2.4	F	G	2.4	1	0.8	90–150	18–22	14	10.5–19.0
	F	A	0	1	0.8–1.2	110–135	18–23	14	9.7–12.7
3.2	F,V,H,O	G	3.2	1	0.8	120–150	20–24	14	10.2–12.7
	F,V,H	A	2.4	1	0.8–1.2	110–135	19–23	14	7.6–11.8
4.8	F,V,H,O	G	4.6	2	0.8–1.2	130–175	22–26	16	10.3–12.7
	F,V,H	B	1.6	1	1.2	140–180	23–27	16	10.3–12.7
	F,V,H	F	1.6	2	1.2	140–175	23–27	16	10.3–12.7
	O	F	1.6	2	1.2–1.6	140–185	23–27	28	10.3–12.7
	F,V	H	2.4–4.8	3	1.2	130–175	23–27	16	10.3–12.7
	H,O	H	4.8	2	1.2–1.6	175–200	24–28	28	10.5–14.8
6.4	F	B	2.4	2	1.2–1.6	185–225	24–29	19	10.3–12.7
	F	F	2.4	3F, 1R	1.2	165–190	25–29	19	10.5–14.8
	V, H	F	2.4	3F, 1R	1.2–1.6	180–200	25–29	21	10.5–14.8
	O	F	2.4	4–6	1.2–1.6	170–200	25–29	28	10.5–16.9
	O, H	H	6.4	1F, 1R	1.6	225–290	26–29	28	8.5–2.7
9.6	F	C–90°	2.4	2F, 1R	1.6	210–275	26–29	24	10.3–14.8
	F	F	2.4	3F, 1R	1.6	190–220	26–29	24	10.3–12.7
	V, H	F	2.4	5F, 1R	1.6	200–250	26–29	26	10.5–16.9
	O	F	2.4	4	1.6	210–290	26–29	38	10.3–12.7
	F, V	H	6.4–9.6	8–10	1.6	190–260	26–29	24	10.5–16.9
	O, H	H	9.6		1.6		26–29	38	
19.0	F	C–90°	2.3	3F, 1R	1.6–2.4	340–400	26–31	28	5.9–8.5
	V,H,O	F	1.6	8F, 1R	1.6	240–300	26–31	38	10.3–12.7
	F	E	1.6	3F, 3R	1.6	270–330	26–31	28	6.8–10.3
	V,H,O	E	1.6	6F, 6R	1.6	230–280	26–31	38	

F : Flat; V : Vertical; H : Horizontal; O : Overhead

Table 12.3 Typical GMAW parameters for fillet welds in aluminum welds with argon shielding

Section thickness (mm)	Welding position	No. of passes	Electrode diameter (mm)	Welding current (DCEP), (A)	Arc voltage (V)	Shielding Gas flow rate (L/min)	Travel Speed (mm/s)
2.4	F, V, H, O	1	0.8	100–130	18–22	14	10–13
3.2	F	1	0.8–1.2	125–150	20–24	14	10–13
	V, H	1	0.8	110–130	19–23	14	10–13
	O	1	0.8–1.2	115–140	20–24	19	10–13
4.8	F	1	1.2	180–210	22–26	14	10–13
	V, H	1	0.8–1.2	130–175	21–25	16	10–13
	O	1	0.8–1.2	130–190	22–26	21	10–13
6.4	F	1	1.2–1.6	170–240	24–28	19	10–13
	V, H	1	1.2	170–210	23–27	21	10–13
	O	1	1.2–1.6	190–220	24–28	28	10–13
9.6	F	1	1.6	240–300	26–29	24	8–11
	H, V	3	1.6	190–240	24–27	28	10–13
	O	3	1.6	200–240	25–28	31	10–13
19.0	F	4	2.4	360–380	26–30	28	8–11
	H, V	4–6	1.6	260–310	25–29	33	10–13
	O	10	1.6	275–310	25–29	40	10–13

F : Flat; V : Vertical; H : Horizontal; O : Overhead

12.1   Striking the Arc

Striking an arc is different with a CC machine than it is with a CV machine. A *CC machine* can usually be set to the desired current by noting the calibrations on the power source, but it is difficult to determine the right wire speed unless an arc is struck. Generally a higher wire speed is deliberately set and then it is reduced progressively until an arc is established and finally set to the required length. Within reasonable limits, adjusting arc length will not affect the welding current.

The welding gun is brought to within 25 mm of the work piece with electrode protruding out about 12 mm from the nozzle. At this stage the gun trigger is pulled and the electrode is contacted with the joint in a slight swinging motion to initiate the arc. Electrode wire feed then begins automatically at the set speed.

Setting a CV arc means first selecting a voltage at the power source and an approximate wire speed setting. Then when the arc is struck, the current can be read from a meter and the wire speed adjusted to give the required current. This will change the arc length to some extent, hence a readjustment of the power source setting may become necessary. This may in turn require a second readjustment of the wire feed speed before the proper combination of current and arc length is achieved. With a CV machine once the trigger is pulled the wire starts feeding immediately. The power source gives a very high current at the instant of short circuit, i.e., when the electrode hits the work. The high current surge tends to explode the tip of the electrode and establish an arc quite readily. Even with a CV power source it is a good practice to use a slight swinging motion of the gun to strike the arc [1201].

Because the electrode wire is fed immediately in GMAW, the arc is established and there is no time for the arc to preheat the base metal. The result is that penetration is very shallow at the weld starts and then gradually increases to the desired depth after about a travel of 25 mm. It can be avoided by striking the arc about 25 mm ahead of the start position and quickly dragging it back to the start position before beginning to weld forward at the normal speed. However it tends to develop porosity at the weld start.

12.2   Welding and Operating Variables

After the arc has been started, the welder has to hold the gun at the correct angles with the desired arc length and traverse the joint at the proper speed. Although arc length is affected by arc voltage, there is no absolute relationship between the two. In practical terms, the welder has to set the correct arc lenght mostly by sound. *The arc should have a light crackling sound.* A too long arc produces a humming sound whereas a too short a arc (which is common to steel) gives rise to a intense crackling and produces unnecessary spatter. A short arc penetrates more than a long arc and for this reason the root pass in a groove weld is made with a somewhat shorter arc than the filling passes. Also, a short arc is used for fillet welding to be certain of penetrating to the root.

Operating Variable Selection

Proper selection of operating variables becomes more difficult because of the interdependence of several of the variables. The practical way of selection of these parameters is to carry-out trial runs to determine an acceptable set of conditions. Typical values of these operating variables for different base metals are listed in Table 12.4.

Table 12.4   Typical conditions for gas metal arc welding of aluminum in the flat position

Material thickness (mm)	Type of weld	Wire diameter (mm)	Current voltage* (A)	(V)	Wire feed speed (mm/s)	Shielding gas	Gas flow (LPM)
1.6	Butt	0.8	90	18	155	Argon	14
3.2	Butt	0.8	125	20	186	Argon	14
5.0	Butt	1.1	160	23	116	Argon	16
6.0	Butt	1.1	205	24	142	Argon	16
10.0	Butt	1.6	240	26	91	Argon	19

*Direct current electrode positive

Weld Reinforcement and Heat Input Control

In recent years there is a significant growth in the use of mechanised weld systems. GMAW offers itself very effectively to such mechanisations. However the process control remain limited to those factors which the machine builders are accustomed, such as electrode feed speed, welding speed, current and voltage. Whereas an welding engineer would prefer to control parameters like reinforcement area, weld heat input, macro and microstructure and mechanical and physical properties of the weld. To achieve this, a model of GMAW process is required that supports direct, independent control of heat and mass input to the weld.

Earlier efforts at modeling the process have concentrated on the electrode melting rate. Lesnewich [1201] proposed an empirical relation for determining electrode melting rate m as a function of current I and the electrode extension L_e, taken equal to contact tube-to-workpiece distance L_{ct} for mild steel with electrode positive as

$$m = (C_1 + C_2 A)I + \frac{C_3 L_e I^2}{(A)^i} \qquad (12.1)$$

where A_e = electrode cross-sectional area and C, C_1, C_2 and C_3 are constants.

Waszink and Van den Henvel [1202] gave an expression of similar form for spray transfer of mild steel

$$m = C_4 I + \frac{C_5 L_e I^2}{A} \qquad (12.2)$$

where C_4 and C_5 are constants.

In both these cases, the electrode melting rate has been given as a function of length of electrode extension L_e. Length of electrode extension is not an easily measurable parameter. Thus, using such expressions for real - time control of the welding process will be difficult. Hence the expression for electrode melting rate needs to be derived in terms of such parameters which can easily be measured and controlled independently. Now it is to be noted that changes in either electrode feed speed or welding speed will change both the heat and mass input to the weld. Therefore, independent control of heat and mass inputs require combined control of both electrode feed speed and weld speed in some unique ratio. Smartt et al. [1203] proposed a process model to calculate the parameters required to obtain desired values of heat and mass input to the weld. The power required to melt the electrode is given by,

$$IV_e + \eta_e IV_a = \frac{\rho_e S_{wf} Q_m d^2 \pi}{4 S_w},$$ (12.3)

where
V_e = Voltage drop across the electrode stickout
V_a = Voltage drop across the arc column
η_e = Efficiency of heat transfer from arc to the electrode
ρ_e = Density of electrode metal
S_{wf} = Electrode feed speed
d = Electrode diameter
S_w = Welding speed
Q_m = Total heat required to melt a unit volume of material and is given as

$$Q_m = \int_{T_0}^{T_s} C_p dT + Q_f$$ (12.4)

where C_p = Specific heat
T_o = Ambient temperature
T_s = Superheat temperature
Q_f = Latent heat of fusion

Now the heat input to the base metal per unit length is given by [1204]

$$H = \eta \frac{VI}{S_w},$$ (12.5)

where V = Arc voltage
η = Heat transfer efficiency from the arc to the base metal
S_w = Welding speed

Weld bead reinforcement G, expressed as the transverse cross-sectional area of the deposited metal, is given by

$$G = \frac{S_{wf} d^2 \pi}{4 S_w}$$ (12.6)

Output voltage of the power supply is given by

$$V = V_0 + nI,$$ (12.7)

where V_o = open circuit voltage

n = effective slope of the $V - A$ curve of the power supply

Power consumed by the process is approximated as the sum of that consumed by resistive heating of the electrode and that consumed in the arc as,

$$IV = IV_e + IV_a, \tag{12.8}$$

Now
$$IV_a = I^2 R_a, \tag{12.9}$$

where R_a = arc resistance.

In normal GMAW, current is controlled by changing electrode speed, and reinforcement is controlled by changing welding speed. Thus heat and mass input to the weld are not controlled independently. Wire feed speed and welding speed can be expressed as

$$S_{wf} = \eta \frac{4GI(V_o + nI)}{\pi d^2 H} \tag{12.10}$$

and
$$S_w = \frac{\pi d^2}{4G} S_{wf} \tag{12.11}$$

Welding System

The welding current is measured during welding by obtaining the voltage drop across a calibrated shunt placed in the secondary loop. This signal being very noisy, a filter is used based on digital, linear difference technique [1203]. The difference between the measured and calculated current is used as the error signal input $e(t)$ to a proportional-integral *PI* controller [1203] which calculates a change in the parameter L_{ct} corresponding to the contact tube-to-workpiece distance as follows:

$$\Delta(L_{ct}) = K_p e(t)$$

$$= K_i \int_o^t e(t)\, dt \tag{12.12}$$

where K_p and K_i are controller proportional and integral gains.

Thus the value of L_{ct} is continuously adjusted in equation (12.12) to reduce the difference between the measured and calculated current. However, the actual contact tube-to-workpiece distance is not changed. In actual operation, desired values of weld reinforcement and heat input per unit length of weld are set on the operator's panel. The system then controls electrode feed speed and welding speed based on model solutions and current feedback, to give the desired values. The control system block diagram is shown in Fig. 12.1.

12.3 Stopping the Arc

Once the trigger is released, the arc extinguishes and no more filler metal can be added to the weld pool to fill the crater. Consequently, without special precautions,

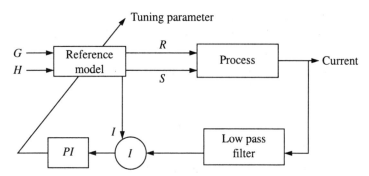

Fig. 12.1   GMAW control system block diagram

a crater will result and it will almost always be cracked. The formation of such craters can be avoided by using run-on/run-off tabs. Alternatively, the weld pool size has to be reduced to a minimum at the moment of extinguishing the arc so that a very little metal contraction takes place on freezing. This will not result in a significant formation of crater and hence cracking can be avoided. The most practical method to achieve this is to accelerate the arc travel speed just before releasing the trigger.

Gas Tungsten Arc Welding

The base metal is melted by an AC or DC arc between a tungsten or tungsten alloy electrode and the base metal itself. A separate filler metal is used. It is not melted by the arc, but by the molten base metal. If the arc contacts the filler, the electrode becomes contaminated by attraction forces between the arc and filler metal. The electrode is shielded by an inert gas that flows through the nozzle in the GTAW torch. Usually argon, helium or mixture of the two is used as shielding gas. The weld metal may be composed of base metal or a mixture of base and filler metal. The welding may be carried out with or without a filler metal. Thicknesses of aluminum alloys commonly welded by GTAW range from 1 mm to 10 mm for manual welding.

REFERENCES

1201   Lesnewich, A. 1958, "Control of melting rate and metal transfer in gas-shielded metal-arc welding. Part II—Control of metal transfer", *Welding Journal* 37(4), pp. 418–s to 425–s.

1202   Waszink, J.H., and Van den Heuvel, G.J.P.M. 1982, "Heat generation and heat flow in the filler metal in GMA welding" *Welding Journal* 61(8), pp. 269–s to 282–s.

1203   Smart, H.B., and Einerson, C.J., "A model heat and mass input control in GMAW", *Welding Journal,* Nov. 1993, pp. 217–s to 229–s.

1204   Rosenthal, D. 1946, "The theory of moving sources of heat and its application to metal treatments", *Trans. ASME*, pp. 849–866.

13

Power Supply

For joining aluminum alloys, GTAW uses either alternating or direct current. Single phase alternating current welding transformers should have an open-circum voltage of 80 to 100 V.

Alternating Current

Usually, alternating current welding provides the optimum combination of current-carrying capacity, arc controllability, and arc cleaning action for the welding of aluminum alloys. A short arc length must be maintained to obtain sufficient penetration and to prevent under cutting, excessive width of weld bead, and consequently loss of control of penetration and weld contour.

Arc length should be about equal to the diameter of the tungsten electrode. On fillet welds, a short arc and adequate current are needed to prevent bridging the root. A short arc also ensures that the inert gas completely surrounds the weld as it forms. The oxide removal action takes place only during the electrode-positive cycle. To ensure this half cycle, the power source must either have a high enough open-circuit voltage (125 volts rms for argon) or a high voltage at high frequency must be impressed on the arc gap at the time that the current passes through zero and the electrode becomes positive.

Using high frequency voltage the arc can be started without touching the electrode to the work. A stable arc (one with steady uninterrupted current flow in each direction) is characterised by the absence of a snapping or a cracking sound, a smooth flow of filler metal into the molten pool of weld metal, easy arc starting and elimination of tungsten inclusions in the weld.

14

Electrodes for Gas Tungsten Arc Welding

For alternating current GTAW, unalloyed tungsten or tungsten zirconiated electrodes are used. The zirconiated electrodes are less likely to be contaminated by aluminum and have a slightly higher current rating. Thoriated tungsten electrodes are not recommended for welding with AC. Fine tungsten inclusions result when using thoria-tungsten electrodes with AC power. When welding aluminum by GTAW with alternating current, the tip of the electrode should be hemispherical as shown in Fig. 14.1. When an electrode becomes contaminated with aluminum, it must be replaced or cleaned. Minor contamination can be burned off by increasing the current while holding the arc on a piece of scrap metal. Severe contamination can be removed by grinding or by breaking off the contaminated portion of the electrode contour on a piece of scrap aluminum.

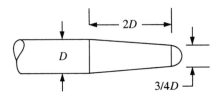

Fig. 14.1   Tungsten electrode tip shape for aluminum welding with AC power

Classification of Electrodes

The classification of tungsten electrode is done on the basis chemical composition, as summarized in Table 14.1 [1401].

An important parameter for characterising tungsten electrodes is the work function of the electrode material. The work function of a material indicates the amount of energy required to give off electrons i.e. to allow current to flow. This is very much relevant to GTAW since the lower the work function of an electrode, the lower the voltage necessary to strike an arc, thus it becomes easier to start the arc.

The work function of pure tungsten is 4.4 eV [1402, 1403]. The addition of

Table 14.1 **Colour code and alloying elements for various tungsten electrode alloys**

AWS classification	Colour	Alloying element	Alloying oxide	Nominal Weight of alloying Oxide (%)
EWP	Green	–	–	–
EWCe-2	Orange	Cerium	CeO_2	2
EWLa-1	Black	Lanthanum	La_2O_3	1
EWTh-1	Yellow	Thorium	ThO_2	1
EWTh-2	Red	Thorium	ThO_2	2
EWZr-1	Brown	Zirconium	ZrO_2	0.25
EWG	Gray	Not Specified		

a stable oxide of a metal, having a work function lower than that of pure tungsten, lowers the work function of the tungsten alloy [1404], i.e. makes it easier to give off electrons and strike an arc. The work function of some of the elements whose oxides are used as additions to tungsten electrodes are given in Table 14.2 [1402, 1403].

Table 14.2 **Work function of various elements**

Elements	Work function (eV)
Zirconium	4.2
Thorium	3.4
Lanthanum	3.3
Cerium	2.6

The addition of these oxides reduces the work function of tungsten electrodes. The oxide additions also affect the current-carrying capacity of the electrodes. This is because, pure tungsten is a metallic conductor while the oxides are insulators. Thus, the electrode with the lowest volume of oxides should have the highest current-carrying capacity, all other factors being equal. Thus tungsten electrodes alloyed with higher density oxides can handle higher welding currents [1405] than pure tungsten electrodes. The alloying additions typically improve arc starting characteristics, help stabilize the arc and increase the life of the electrodes [Ref. 1405]. The various types of electrodes have distinct electrical characteristics that result in different welding performances and weld bead shapes. In all welding applications where precise parameter control is used, changing from one classification of electrode to another may require procedure redevelopment and requalification.

Pure Tungsten Electrode (EWP)

Electrodes containing minimum 99.5% by weight tungsten with no internal alloying elements are classified as pure tungsten electrodes. The current carrying capacity of these electrodes is lower than that of the alloyed electrodes. These are mainly used with alternating current for welding of aluminum and magnesium alloys. With this type of current, the tip of the EWP electrode forms a clean balled end, which provides good arc stability. Direct-current-electrode-positive welding does not provide arc initiation and arc stability and may also cause splitting and melting of pure tungsten electrodes.

Electrode Sizes and Current Capacities

Tungsten electrodes are available in a variety of standard diameters from 0.30 to 8.00 mm. The choice of an electrode, its size and welding current depend on the type and thickness of base metals being welded. The current-carrying capacities of all types of tungsten electrodes are affected by the type of welding torch, the type of power source (DCEN, DCEP, AC), the electrode extension beyond the collect, and the shielding gas.

The suggested current ranges for each electrode type and size using argon shielding gas along with gas cup diameters recommended for use with different types of welding power sources are listed in Tables 14.3 and 14.4 [1406]. In an arc, about 70% of the heat is generated at the anode and 30% at the cathode. Therefore with electrode negative i.e. cathode, it can carry a much higher current without over-heating than in case of electrode positive polarity. Whereas, direct current with the electrode positive requires a much larger diameter to support a given level of current because the tip is not cooled by the emission (evaporation) of electrons but is heated by their impact.

In case of AC power supply, the electrode tip is cooled during each electrode negative half cycle and is heated in the positive cycle. Hence the current-carrying capacity of an electrode with AC supply is between that of DECEN and DCEP. Therefore, a larger diameter electrode is required in case of welding with AC supply for a given current level than does DCEN. Operating at higher current levels than recommended for a given electrode size may cause the electrode to erode, melt or split. Also tungsten particles may get into the weld pool and act as discontinuities or defect in the weld. Whereas operating at too low a current for a specific electrode diameter may cause arc instability. In general use of the smallest recommended diameter for a given current level is usually the best [1404].

Electrode Tip Configuration

The shape of the electrode tip is an important process variable in GTAW. This should be studied during welding procedure development, and adhered to during production, specially for mechanised or automated welding. Regardless of the electrode tip geometry selected, it is imporant that a consistent tip configuration be used once a welding procedure is established. Table 14.5 is a guide for electrode tip preparation for a range of sizes with recommended current ranges. The angle of the cone, which is formed at the end of a tungsten electrode, affects the penetration shape of the weld metal. A small cone angle can produce a plasma jet giving deep narrow penetration when welding parameters are such that the full thickness of the work-piece is penetrated. On thicker plates where there is a partial penetration, the plasma jet gets deflected causing a wide bead at the surface. With a large cone angle, the plasma jet does not develop and

Table 14.3 Typical type of current, tungsten electrodes and shielding gases for welding different metals

Type of metal	Thickness (mm)	Type of current	Electrode	Shielding gas
Aluminum	All	AC	Pure or zirconium	Argon or argon-helium
	over 3	DCEN	Thoriated	Argon-helium or argon
	under 3	DCEP	Thoriated or zirconium	Argon
Copper, Copper alloys	All	DCEN	Thoriated	Helium or argon-helium
	under 3	AC	Pure or zirconium	Argon
Magnesium alloys	All	AC	Pure or zirconium	Argon
	under 3	DCEP	Zirconium or Thoriated	Argon
Nickel, Nickel alloys	All	DCEN	Thoriated	Argon or argon-helium
Plain carbon, Low-alloy steels	All	DCEN	Pure or zirconium	Argon
	under 3	AC	Thoriated	Argon
Stainless steel	All	DCEN	Thoriated	Argon or argon-helium
	under 3	AC	Pure or zirconium	Argon
Titanium	All	DCEN	Thoriated	Argon

Thoriated electrodes are recommended, ceriated or lanthanated electrodes may also be used.
AC is recommended, variable polarity or pulsed current can be used.

Table 14.4 Typical current ranges for tungsten electrode for GTAW

Electrode diameter (mm)	Gas cup ID (mm)	Direct Current (Amps)		Alternating current (amps)	
		DCEN	DCEP	Unbalanced wave AC	Balanced wave AC
		EWP EWCe-2 EWLa-1 EWTh-1 EWTh-2	EWP EWCe-2 EWLa-1 EWTh-1 EWTH-2	EWP / EWCe-2 EWLa-1 EWTh-1 EWTh-2 EWZr-1	EWP / EWCe-2 EWLa-1 EWTh-1 EWTh-2 EWZr-1
0.25	6	up to 15	N/A	up to 15	up to 15
0.50	6	5–20	N/A	5–15	10–20
1.00	10	15–80	N/A	10–60	20–30
1.50	10	70–150	10–20	50–100	30–80
2.50	12	150–250	15–30	100–160	60–130
3.175	12	250–400	25–40	150–210	100–18C
4.000	12	400–500	40–55	200–275	160–240
4.750	15	500–750	55–80	250–350	190–300
6.350	18	750–1000	80–125	325–450	250–400

Unbalanced wave AC (EWCe-2 EWLa-1 EWTh-1 EWTh-2 EWZr-1): up to 15, 5–20, 15–80, 70–150, 140–235, 225–325, 300–400, 400–500, 500–630

Balanced wave AC (EWCe-2 EWLa-1 EWTh-1 EWTh-2 EWZr-1): up to 15, 5–20, 20–60, 60–120, 100–180, 160–250, 200–320, 290–390, 340–325

Table 14.5 Examples of tungsten electrode tip shapes and current ranges

DCEN (Electrode negative/straight polarity with argon as shielding gas)				
Electrode diameter (mm)	Diameter at tip (mm)	Constant included angle (degrees)	Current range (A)	Pulsed current range (A)
1.02	0.125	12	2–15	2–25
1.02	0.25	20	5–30	5–60
1.59	0.5	25	8–50	8–100
1.59	0.8	30	10–70	10–140
2.38	0.8	35	12–90	12–180
2.38	1.1	45	15–150	15–250
3.18	1.1	60	20–200	20–300
3.18	1.5	90	25–250	25–350

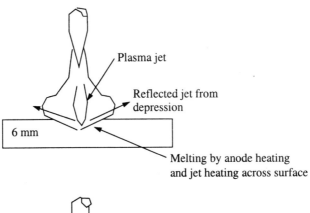

Plasma jet

Reflected jet from depression

6 mm

Melting by anode heating and jet heating across surface

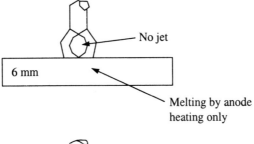

No jet

6 mm

Melting by anode heating only

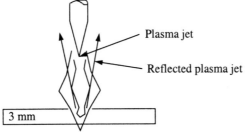

Plasma jet

Reflected plasma jet

3 mm

Fig. 14.2   Effect of electrode cone angle for tungsten-arc welding

hence thick plates can be welded with a narrow bead. These apparently contradictory effects of electrode tip shape are shown in Fig. 14.2.

REFERENCES

1401 ANSI/AWS A5 12–92, "Specification for Tungsten and Tungsten Alloy Electrodes for Arc Welding and Cutting", *American Welding Society,* Miami, 1992.
1402 R.C. Weast, Ed., *Handbook of Chemistry and Physics,* 1973–1974, CRC Press, 54th Ed. pp. E–82 to E–82.
1403 Osram Sylvania, Inc. Literature.
1404 Campbell, R.D. and Lacoursire E.J., "A Guide to the Use of Tungsten Electrodes for GTA Welding", *Welding Journal,* Jan. 1995, pp. 39–45.
1405 Ushio, M. and Matsuda, F., 1986. "Study on gas tungsten arc electrodes: Comparative study on characteristics of tungsten-oxide cathode", *IIW Document Number 212-648-86.* International Institute of Welding.
1406 *Welding Handbook,* 8th Edition, Vol. 2, 1991, American Welding Society, Miami.

15

Shielding Gases for Tungsten Arc Welding

15.1 Argon

The most commonly used shielding gas is argon. It is heavier than air and its density is 1.7837 g/l at standard temperature and pressure. The welding grade argon is refined to a minimum purity of 99.9995%. It is chemically inert, colourless, odorless, tasteless and nontoxic gas.

The prime factor that influences shielding effectiveness is the gas density. Argon is approximately 1.3 times heavier than air and 10 times heavier than helium. After leaving the gas nozzle, it tends to form a blanket over the weld area. Argon is used for welding of a wide range of materials such as, mild steel, aluminum, copper, stainless steel, nickel alloys and the reactive metals like titanium and magnesium.

Argon is used more extensively than helium because of the following advantages:

- Easier arc initiation
- Better control of the molten weld pool
- Smoother and quieter arc action
- Lower cost and greater availability
- Better cross-draft resistance due to higher density
- Lower flow rates for shielding
- Cleaning action when welding materials such as aluminum and magnesium with AC or DCEP
- Reduced penetration

15.2 Helium

Helium being one of the lighter gases, it tends to rise around the gas nozzle. Hence, to achieve effective shielding the flow of helium must be 2 to 3 times more than that of argon. It has a higher ionization potential, approximately 25 eV compared to 16 eV for argon. For given values of welding current and arc length, helium tansfers more heat into the work than argon. Since the arc in the helium shield is considerably hotter than with argon and because it has a higher thermal conductivity, it can achieve higher welding speeds and improve the weld bead penetration profile. The grater heating power of the helium arc is advantageous for joining of metals of high thermal conductivity and for high speed mechanized applications.

When using DCEN for automated GTAW of aluminum, helium shielding provides a greater fusion depth and higher welding speeds than argon. However when using helium as a shielding gas for GTA welding of aluminum with DCEN power supply, mechanical removal of surface oxides becomes necessary. Although helium offers certain advantages, its major drawbacks are:

- Produces less stable arc.
- Arc initiation is difficult.
- Much more expensive than argon.
- Less availability.

15.3   Argon and Helium

Helium-argon gas mixture is used to take advantages of the optimum operating characteristics of each gas. The superior arc initiation and stable arc characteristics of argon and helium's higher thermal conductivity produce high-quality GTA welding on aluminum when using AC. Increasing helium content, the travel speed as well as the depth of fusion for both manual and mechanical welding of nonferrous alloys increase. The helium content usually ranges between 25 and 75%.

15.4   Shielding Gas Flow Rates

Shielding gas flow rates are based on gas nozzle size, weld pool size and the air movement. The gas nozzle diameter is selected to match the expected size of the weld pool and the reactivity of the metals to be welded. The minimum flow rate should be such that it will produce a stiff stream of shielding gas to overcome the heating effects of the arc and local cross drafts of air.

Typical flow rates with manual torches are 5 to 12 liters/min for argon and 12 to 21 liters/min for helium. Excessive flow of shielding gas can be harmful. It can give rise to turbulence in the gas stream, which may take in atmospheric contamination into the weld pool. It may also cause instability in the welding arc. An air-draft of approximately 8 to 10 km/h can disrupt the shielding gas coverage. Protective screens are used to prevent interference by the wind.

Characteristic of Argon and Helium
The important characteristics of these gases are the volt-ampere relationships of the tungsten arc in argon and in helium as shown in Fig. 15.1. The arc voltage with helium as shielding gas is always higher for equivalent arc lengths over the entire current range. Therefore helium offers more heat than argon, since the arc power which is the product of voltage and current is higher in helium as compared to argon. Hence when welding thick materials and metals having high thermal

conductivity or relatively high melting temperature, helium is preferred to argon as a shielding gas.

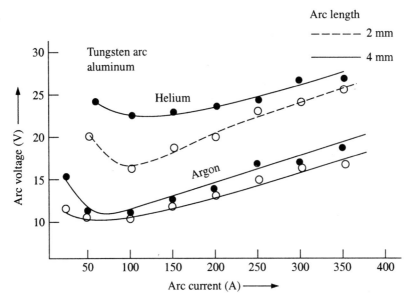

Fig. 15.1 Voltage-current relationship with argon and helium shielding

At lower currents, the volt-ampere curves pass through a minimum voltage. At around 90 ampere, the voltage is minimum and then with decreasing current the voltage increases. For helium, this increase in voltage occurs in the range of 50 to 150 A. Since the voltage increase for argon occurs below 50A, the use of argon in the 50 to 150 A range provides the welder with more flexibility in arc lengths to control the welding operation. Both gases provide excellent arc stability with direct current power. However with AC power argon provides much better arc stability and the cleaning action of the weld pool and this makes argon superior to helium in this respect. Because of this, AC power with argon as shielding gas is extensibly used for welding of aluminum and magnesium alloys. *Argon is used as a shielding gas more often than helium,* because it:

- Is more readily available and costs less than helium
- Affords better control of the weld pool
- Gives a smoother, quieter arc, greater arc cleaning action, and easier arc starting
- Requires less gas for specific applications
- Has better cross-draft resistance than helium
- Causes less clouding and the metal stays brighter; the operator can thus see the weld pool more easily

Argon is preferred for direct current electrode positive welding because it establishes an arc more easily and provides better arc control. When aluminum is welded with alternating current or DCEP, a white band of varying width appears alongside the weld bead.

This white band, disclosed by one analysis to be aluminium oxide, never occurs when welding is done with DCEN, and it is believed to be caused by the emission of electrons from the surface of the aluminum when it is on the cathode side of the arc (DCEP). The electrons leave the aluminum through the aluminium oxide, thus serving to detach the oxide from the surface.

16

Welding Procedures

Typical welding schedules for manual alternating current GTAW are given in Tables 16.1 and 16.2. Metal that is more than 3 mm thick is grooved to ensure complete penetration. A single V-groove with an include angle of 60 to 90° is used most often. If the operator has difficulty in maintaining a very short arc, a larger included angle (a wider edge preparation) or joint spacing may be required to allow a longer arc to be used.

16.1   Starting the Arc

The arc is struck like striking a match by swinging the electrode holder in a pendulum like motion toward the starting place. With superimposed high-frequency current for arc stabilization, there is no need to touch the work with the electrode because the arc starts when the electrode tip is brought close to the work surface. A short arc length must be maintained to obtain sufficient penetration and to avoid undercutting, excessive width of the weld bead, and consequent loss of control of penetration and weld contour. Approximately the arc length should be equal to the diameter of the tungsten electrode. The gas nozzle must be as small as possible while still providing adequate shielding of the weld pool. The torch should be vertical to the centreline of the joint, at a forehand angle of about 75 to 85° from the plane of the job. When welding unequal sections, the arc should be directed towards the heavier section.

16.2   Welding and Operating Variables

The arc is held at the starting point until the metal liquefies and a weld pool is established. Establishment and maintenance of a suitable weld pool is important, and welding must not proceed ahead of the pool. When using correct current, the travel speed is higher leading to less dissipation of heat. This in turn promotes progressive solidification of the weld metal and results in better weld bead control.

When the arc is broken shrinkage cracks may occur in the weld resulting in a defective weld. This can be avoided by gradually lengthening the arc several times while adding more filler metal into the crater, or by using a foot control to reduce the current while adding filler at the end of a weld.

Table 16.1   Chemical treatments for oxide removal prior to welding or brazing aluminum

Type of solution	Concentration	Temperature	Type of container	Procedure	Purpose
Sodium hydroxide	NaOH 1.76 oz (50 grms). 26 per gal (1 L) of water	60-71° C	Mild steel	Immerse for 10 to 60 sec. Rinse in cold water	Removes thick oxide for all welding and brazing procedures. Active etchant.
Nitric acid	Equal parts of HNO_3 (68 %) and Water	Room	Type 347 Stainless steel	Immerse for 30 sec. Rinse in cold water. Rinse in hot water and dry	
Sulfuric acid	H_2SO_4 1 gal (3.79 L) CrO_3—45 oz (1.28 kg) water 9 gal (34. 1L)	60°-82° C	Antimonal Lead lined steel tank	Dip for 2-3 min. Rinse in cold water. Rinse in hot water and dry	For removal of heat treatment and annealing films and stains and for stripping oxide coatings.
Phosphoric-chromic	H_3PO_4 (75%) – 3.5 Gel (13.3 L) $CrO3$ 1.75 lb (79.4 grams) water—100 gal (379 L)	93°C	Type 347 Stainless steel	Dip for 5-10 min. Rinse in cold water. Rinse in hot water and dry	For removal of heat treatment and annealing films and stains and for stripping oxide coatings.
Sulfuric acid	H_2SO_4—5.81 oz (165 grams) water - 0.26 gal (1 L)	73°C	Polypropylene lined steel tank	Immerse for 5-10 min. Rinse in cold water. Rinse in hot water and dry	Oxide removal. Mild etchant.
Ferrous sulfate	$Fe_2SO_4H_2O$ 10% By volume	26.7°C	Polypropylene	Immerse for 5-10 min. Rinse in cold water. Rinse in hot water and dry	Oxide removal.

Table 16.2 Typical procedures for manual gas tungsten arc welding of butt joints in aluminum with AC and argon shielding

Section thickness (mm)	Welding position	Joint geometry	Root opening (mm)	Number of weld passes	Filler rod diameter (mm)	Electrode diameter (mm)	Welding current (A)	Travel speed (mm/s)
1.6	F, V, H	B	0-1.6	1	1.6, 2.4	1.6	60-80	3.4-4.2
	O	B	0-1.6	1	1.6, 2.4	1.6	60-75	3.4-4.2
2.4	F	B	0-2.4	1	3.2	2.4	95-115	3.4-4.2
	V, H	B	0-2.4	1	2.4, 3.2	2.4	85-110	3.4-4.2
	O	B	0-2.4	1	2.4, 3.2	2.4	90-110	3.4-4.2
3.2	F	B	0-3.2	1-2	3.2, 4.0	2.4	125-150	4.2-5.1
	V, H	B	0-2.4	1-2	3.2	2.4	110-140	4.2-5.1
	O	B	0-2.4	1-2	3.2, 4.0	2.4	115-140	4.2-5.1
4.8	F	D–60°	0-3.2	2	4.0, 4.8	3.2	170-190	4.2-5.1
	V	D–60°	0-2.4	2	4.0	3.2	160-175	4.2-5.1
	H	D–90-°	0-2.4	2	4.0	3.2	155-170	4.2-5.1
	O	D–110°	0-2.4	2	4.0	3.2	165-180	4.2-5.1
6.4	F	D–60°	0-3.2	2	4.8	4.0	220-275	3.4-4.2
	V	D–60°	0-2.4	2	4.8	4.0	200-240	3.4-4.2
	H	D–90°	0-2.4	2-3	4.0, 4.8	4.0	190-225	3.4-4.2
	O	D–110°	0-2.4	2	4.8	4.0	210-250	3.4-4.2
9.5	F	D–60°	0-3.2	2	4.8, 6.4	4.8	315-375	3.4-4.2
	F	E	0-2.4	2	4.8, 6.4	4.8	340-380	3.4-4.2
	V	D–60°	0-2.4	3	4.8	4.8	260-300	3.4-4.2
	V, H, O	E	0-2.4	2	4.8	4.8	240-300	3.4-4.2
	H	D–90°	0-2.4	3	4.8	4.8	240-300	3.4-4.2
	O	D–110°	0-2.4	3	4.8	4.8	260-300	3.4-4.2

The filler metal is to be added to the leading edge of the weld pool but to one side of the centre line. The tungsten electrode should not touch the filler rod, and the end of the filler rod should not be withdrawn from the argon shield.

If the tungsten electrode gets contaminated with aluminum, it must be replaced or cleaned. Increasing the current, while holding the arc on some scrap metal, can burn off minor contamination. Severe contamination is to removed by grinding or by breaking the contaminated part and reforming the correct electrode contour on a piece of scrap aluminum. Tacking of plates before welding is to be done to prevent distortion. The tack welds should be of ample size and strength to withstand the forces due to welding thermal cycle. The tacks should be chipped out or tapered at the ends before welding over them.

Operating Variables
The principal operating variables in case of GTAW are:

- Arc voltage (arc length),
- Welding current,
- Welding speed,
- Shielding gas.

All these variables interact with each other very strongly, therefore they can not be treated as independent variables while establishing welding procedures for fabricating specific joints.

Welding Current
Depending on the type of metal to be welded either DC or AC is used. Deep penetration and fast welding speeds can be achieved with DC electrode negative, especially with helium as shielding gas. For welding aluminum, AC is used. It provides a cathodic cleaning (sputtering) of the weld pool and removes the refractory oxides from the weld deposit and results in a superior weld. With AC to have sputtering effect argon has to be used as shielding gas. Sputtering does not take place with helium. In case of manual GTAW always argon is used as a shielding gas irrespective of AC or DC power supply.

Polarity
With DC power supply the tungsten electrode may be connected to either the negative or positive terminal of the power supply. However almost always electrode negative (DCEN) is used and electrode positive (DCEP) is generally limited to welding sheet metal. The characteristics of current polarity for GTAW are shown in Fig. 16.1. With DCEN, approximately 70% of the heat is generated at the plate (anode) and 30% at the electrode (cathode). Therefore for a given current, a deeper penetration is achieved with DCEN as compared to that of DCEP.

With DCEP, a cathodic cleaning action is created at the surface of the work-piece. This action occurs with most metals, however this is of special importance when welding aluminum and magnesium because it removes the refectory oxides that inhibits wetting of weldment by the weld metal. With DCEP, the electrode tip gets overheated by the bombardment of electrodes as well as by its resistance

Current type	DCEN	DCEP	AC (balanced)
Electrode polarity	Negative	Positive	
Electron and ion flow			
Penetration characteristics			
Oxide cleaning action	No	Yes	Yes-once every half cycle
Heat balance in the arc (approx)	70% at work end 30% at electrode	30% at work end 70% at electrode	50% at work end 50% at electrode
Penetration	Deep, narrow	Shallow, wide	Medium
Electrode capacity	Excellent 3.2 mm, 400 A	Poor 6.4 mm, 120 A	Good 3.2 mm, 225 A

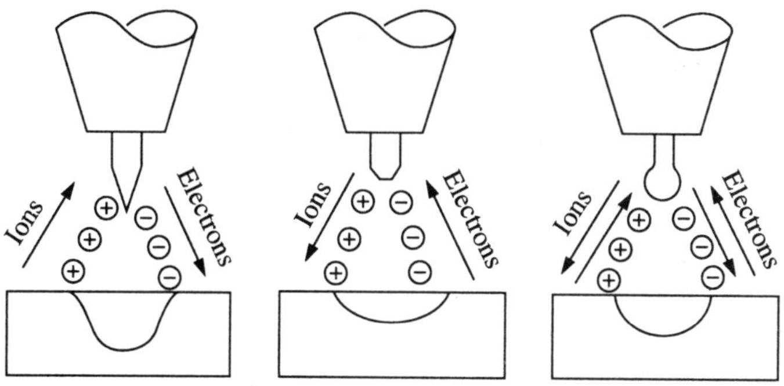

Fig. 16.1   Characteristics of current types for gas tungsten arc welding

to their passage through the electrode. Therefore to reduce resistance heating as well as to provide better thermal conduction, a larger diameter electrode is used with DCEP welding. The current carrying capacity of an electrode with DCEP is approximately 1/10th that of an electrode connected to the negative terminal i.e. DCEN.

With alternating current power source, the cleaning action of DCEP and deep penetration of DCEN both are achieved. With conventional AC power sources, some means of stabilizing the arc during voltage reversal is required. When the current decays to zero with the electrode becoming negative, it supplies electrons immediately to re-ignite the arc. Whereas when the weld pool becomes negative, it can not supply electrons until the voltage is raised sufficiently to initiate cold-cathode emission. Without this voltage, the arc becomes unstable.

The arc stabilization can be achieved by one of the following ways:

- High open circuit voltage
- By discharging capacitors at appropriate time in the cycle
- High-voltage, high-frequency sparks in parallel with the arc
- Using square wave output

Arc Voltage

Arc voltage is used in describing welding procedure mainly because it is easy to measure. Arc voltage is very much dependent on other variables [1601]

- Welding current

- Shape of tungsten electrode tip
- Distance between electrode tip and workpiece
- Type of shielding gas.

An increase in the arc length will generate an increase in the arc voltage. Thus arc voltage becomes a way to control the arc length. Arc length is an important parameter in this process, however it is very difficult to monitor. The magnitude of heat flux is proportional to the arc voltage [1602]. However, the distribution parameter increases nearly linearly with arc length [1602]. Also, the arc efficiency substantially decreases when the arc length increases [1603]. The weld penetration slowly increases with the arc length, till the arc length reaches a critical value where the weld penetration slowly decreases with further increase in arc length. However, beyond a certain arc length again the penetration starts increasing [1604]. Thus the relationship between the weld penetration and the arc length is quite complicated.

It is generally expected that the weld pool width increases with arc length. However, recent studies [Ref. 1604] have shown that, the weld pool width remains almost unchanged before the arc length reaches a certain value. Beyond that value, with increasing arc length, pool width starts decreasing. This happens because, the arc distribution does not keep increasing with arc length. Before the arc length reaches a critical value, the arc spread begins to concentrate. Because of the concentration of the arc, the back side bead width becomes large [1604].

If the welding arc is too short, there is a possibility of short circuiting the electrode to the molten pool. However, with mechanised welding, using helium as shielding gas, and high current DCEN power, it is possible to submerge the electrode tip below the plate surface to produce deeply penetrating but narrow welds at higher speeds. This technique is known as buried arc. In most applications other than those involving sheet metal, the desired arc length should be as short as possible.

Welding Speed
Welding speed influences both penetration and weld pool width. At constant current and arc length, a slower weld speed will supply more heat input to the weldment and provide a higher peak temperature on the surface, whereas a faster one will result in less heat input and a lower peak temperature. That is, a slower speed arc can be regarded as having a higher power density since the distribution of the arc is not significantly affected by the welding speed.

The weld pool gets enlarged with decrease in the welding speed. The weld pool length becomes large and the rear of the pool becomes sharper with decrease in welding speed [1604]. The effect of welding speed is more pronounced on the molten pool dimensions than that on penetration. Welding speed is also an important operation variable from cost point of view. In some applications, the other variables are selected so as to achieve the desired welding speed for the required weld configuration. In mechanised welding, the welding speed is generally fixed and the other variables such as current and voltage are adjusted to obtain the required weld quality. The various types of current, tungsten electrode and shielding gases as recommended for welding different metals are given in Table 16.3.

Table 16.3 Recommended types of current, tungsten electrode and shielding gases for welding different metals

Type of metal	Thickness	Type of current	Electrode	Shielding gas
Aluminium	All over 3 mm under 3 mm	AC DCEN DCEP Thoriated or zirconium	Pure of zirconium Thoriat Argon	Argon or argon-helium. Argon-helium or argon
Copper, Copper alloys	All under 3 mm	DCEN AC zirconium	Thoriated Pure or	Helium argon
Magnesium alloys	All under 3 mm	AC DCEP zirconium or Thoriated	Pure or zirconium	Argon Argon
Nickel, Nickel alloys	All	DCEN	Thoriated	Argon
Plain Carbon, low-alloy steel	All under 3 mm	DCEN AC	Thoriated Pure or zirconium	Argon or argon-helium argon
Stainless steel	All under 3 mm	DCEN AC zirconium	Thoriated Pure or Argon	Argon or argon-helium
Titanium	All	DCEN	Thoriated	Argon

16.3 Stopping the Arc

Breaking the arc also requires special care to prevent the formation of shrinkage cracks in the weld crater. Several techniques are used. The arc can be quickly broken and restruck several times while adding filler metal to the crater; or a foot control can be used to reduce current at the end of the weld; or travel speed can be accelerated to "tail out" the weld before breaking the arc. By using these techniques and adequate fixturing, weld joints can be made manually in aluminum alloys down to 1mm thick, without excessive distortion.

Tack welding before final welding is helpful in controlling distortion. Tack welds should be of ample size and strength and preferably should be chipped out or tapered at the ends before welding over them. The edges to be welded are grooved by machining. All grease, oil, and oxide are removed by solvent cleaning and brushing with a stainless steel wire brush before welding. The welding current and method of depositing the weld metal are selected to avoid undercutting.

REFERENCES

1601 *Welding Handbook,* 8th Edition, Vol. 2, 1991, American Welding Society, Miami.

1602 Tsai, N.S., and Eagar, T.W., 1985. "Distribution of the heat and current fluxes in gas tungsten arcs", *Metallurgical Transactions* B, Vol. 16B, pp. 841–846.

1603 Giedt, W.H., Tallerico. L.N. and Fuerschbach, P.W., 1989, "GTA welding efficiency: Calorimetric and temperature field measurements", *Welding Journal* 68 (1), pp 28-s to 32-s.

1604 Kovacevic, R., Zhang, Y.M., and Li L., "Monitoring of Weld Joint Penetration Based on Weld Pool Geometrical Appearance", *Welding Journal Research Supplement,* Dec. 1996, pp. 317-s.

17

Weld Discontinuities

Discontinuities in aluminum arc welds can occur for a wide variety of reasons. These can be associated with methods of preparing the metal prior to welding or the use of incorrect welding procedures and techniques. These discontinuities are quite similar to those found in welding other metals and include cracks, incomplete penetration, incomplete fusion, porosity and weld inclusions.

Crater Cracks

Molten aluminum shrinks about 6% in volume as it solidifies [1701]. This is twice as much as ferrous metals. As a result, sudden termination of the welding arc usually results in a weld crater as the weld pool solidifies. Cracks often emanate from the centre of the crater. These crater cracks may be small but are very serious, since they usually occur at the end of a weld where the stress concentration is most pronounced. Not only do the cracks extend from the centre of the crater in many directions, similar to a crow's foot, but they will extend internally at the centre of the crater to the full depth of the weld pass.

Crater cracks can be prevented by good welding techniques. A common technique is to 'tail-out' the weld so as the minimize the size of the weld pool before terminating the arc. This can be done by tapering down the welding current or accelerating the weld travel speed, or both. Tailing-out on a flat surface is preferred rather than in a fillet or groove so as to reduce surface tension effects that restrain the solidifying weld metal. Another technique to reduce crater cracking is to rapidly break and restrike the arc several times at the termination point while feeding additional filler metal to fill the crater during the solidifiction process [1702]. Runoff tabs are also used to avoid crater cracks on the finished part. Crater cracking is usually detected visually, but dye penetrants can be used effectively to detect surface cracks. Operators must be trained to look for crater cracks and not to weld over them but to chip them out before continuing the weld. Many internal cracks in multiple pass welds are a result of not following this practice, only melting the top portion of the crater when welding over it. These internal cracks can be determined by X-ray and ultrasonic means, destructive metallographic sections or bend tests.

Weld Cracking

Weld cracking can be placed into two general categories, cold cracks and hot cracks. Cold cracks occur below the solidus temperature of the weld metal and result from inadequate mechanical strength of the deposit to withstand conditions of restraint and contraction of the parts during cooling. This most commonly occurs in the root pass of a groove or fillet weld and usually can be overcome by depositing a larger initial weld pass. A backhand welding technique may be desirable in this instance with the GMAW process rather than the forehand gun angle normally employed with aluminum. The backhand technique will increase the thickness through the throat of the weld.

Hot cracks occur when the metal is above the solidus melting temperature and result from a tensile stress application upon a low-strength, partially solidified segment of the weld or base metal. Hot cracks most commonly occur when welding the heat treatable aluminum alloys. Hot cracking can result from the use of an incorrect filler metal, excessive base alloy dilution of a weld, an improper joint design or excessive joint spacing. Typically, if the weld metal solidifies before the base metal adjacent to the weld, shrinkage of the weld can impose stresses upon partially solidified base metal to create cracking.

When welding the heat treatable alloys, a general rule of thumb would be to employ a filler metal with a solidus melting temperature equal to or below that of the base metal. Alloy 4145 alloy is the preferred arc welding filler metal to minimize heat-affected zone (HAZ) cracking with 2014 base metal because its solidus temperature is most compatible. Filler metal alloy 2319 has been used successfully with mechanized welds and on thin gauges employing high thermal gradients from the weld to the base metal.

High-speed welding procedures and the use of chill bars to produce a sharp thermal gradient can help reduce HAZ cracking in the thick heat treatable alloys. These procedures allow the HAZ to completely solidify and develop some strength before application of the weld shrinkage stresses. Slow, oscillating techniques reduce the thermal gradient and increase the susceptibility to hot cracking in the partially melted HAZ, as well as in the weld metal. With the heat treatable aluminum alloy, the joint bevel,, particularly with the 6XXX series, should allow the weld deposit to consist of 50% or more of filler metal. Alloy 6061, even in a thickness as thin as 2.3 mm, is beveled to permit a minimum of 50% dilution with 4XXX series (AI–Si) filler metals and approximately 75% dilution when using 5356 (AI–5Mg) filler metal [1703]. Application of a compressive loading to the parts during welding can eliminate hot cracks. Typical of this would be a circumferential weld in a small tank, such as a beer barrel, where the two halves of a 6XXX series alloy are positioned in a lathe for welding and an end force applied allows an autogenous GTA weld to be made satisfactorily. In a similar fashion, forge pressure is used during resistance spot welding to apply a compressive load to overcome cracking in most high strength, heat treatable alloys (2XXX and 7XXX series) during solidification of the weld. Radiographic and ultrasonic inspection methods will reveal internal cracks; however, ultrasonic examination is more likely to detect the small intergranular hot cracks. Dye penetrant readily shows surface cracking that may not be observed visually. Metallographic· examination will readily reveal internal or surface cracking.

Incomplete Joint Penetration

Incomplete joint penetration results when the weld bead does not penetrate the full thickness of the base metal when welding from one side, or where the weld beads do not interpenetrate when welding from both sides of a butt joint. Although a serious discontinuity in either case, it is particularly significant in reducing the joint strength when welding is done from one side only.

In groove welds, this discontinuity results from insufficient welding current, an excessive welding speed, an inadequate bevel angle or depth, and insufficient joint spacing. When welding from both sides, the same factors apply, as well as an improper back gouge depth or shape. In fillet welds, incomplete root penetration results when the filler metal does not penetrate into the weld root. In addition to insufficient welding current or excessive welding speed, an excessive arc length can cause poor root penetration in fillet welds. Generally, a short arc is necessary to concentrate the heat at the root of a fillet to obtain proper root fusion and root penetration.

In structural aluminum welding, it is not necessary to have the welds interpenetrate in a two-sided fillet. Adequate fillet size will provide ample strength to equal the strength of the heat-affected zone of the base metal joint without fusing the two fillets together. Particularly in heat-treatable aluminum alloys, the higher heat input necessary to have the two fillet welds interpenetrate would tend to further reduce the strength of the HAZ of the base metal and produce lower strength than a joint with normal but adequate fillet size. Fabrication of single fillet welds and evaluation by facture and metallographic methods, as used for procedure qualification tests, are the easiest means to determine suitability of welding parameters in producing full penetration fillet welds.

Incomplete Fusion

Incomplete fusion is a discontinuity that can be difficult to detect in aluminum welds. It occurs when the refractory oxide film, which is present on the surface of aluminum, is not completely removed by the cleaning action of the arc. This oxide melts at approximately 2035°C and can remain intact on the surface of the metal even though the alloy under the oxide melts at about 650°C or less. Thus, the oxide can act to prevent the melted segments from flowing together into a properly fused weld. This thin membrane of oxide is insoluble in the molten weld pool. It often is left intact when the edges being welded fit tightly together. The more thickly oxidized the aluminum components (as a result of thermal treatments, water stain, anodic treatments, etc.) the greater the tendency to encounter poor fusion in aluminum welds. Mechanical or chemical removal of these thicker oxides is very desirable in obtaining sound welds.

Aluminum oxide has the same radiographic density as aluminum and will not be detected by X-ray. However, as the weld solidifies, it shrinks and a shallow void will remain at the location of the incomplete fusion. It is the void, rather than the oxide inclusion, that is visible by X-ray when the void is properly oriented. Ultrasonic inspection will readily detect the incomplete fusion. Also, incomplete fusion will be very evident in destructively tested specimens.

Basic causes of incomplete fusion when welding aluminum are as follows:

- Welding in a narrow-groove with poor root accessibility.
- Insufficient heat input or excessive travel speed.
- Excessive arc length.
- Improper torch angle impingement or alignment.
- Inadequate gas shielding.
- Improper back gouge depth or shape (U-shape is highly preferred over a V-shape)
- Insufficient breakup of root oxides when welding from one side.

With gas metal arc welding, the use of a constant voltage (CV) type power supply or an arc voltage controlled wire feed system can contribute to intermittent incomplete fusion if the arc is oscillated during the welding operation. A 'drooping' volt/ampere characteristic (constant current) power supply is preferred, when GMAW is manipulated, to obtain most uniform penetration and fusion characteristics in aluminum alloys.

Helium additions (50–75% of total mix) to the argon shielding, when welding with 5XXX series aluminum electrodes, are beneficial in widening the penetration pattern of the weld and working the oxide deposit out of the joint to overcome fusion difficulties [1704, 1705].

Porosity

Porosity or gas holes can be present in aluminum welds to varying degrees depending upon the amount of hydrogen introduced into the weld zone. The solubility of hydrogen in molten aluminum is about 19 times more than it is in solid aluminum. Consequently, if the weld pool has dissolved hydrogen, this gas is rejected from solution during solidification of the weld and, if the weld solidifies sufficiently fast, the gas becomes trapped within the weld deposit. Thus, to overcome the primary cause of porosity in aluminum welds, sources of hydrogen such as moisture and hydrocarbons must be removed from the weld zone. In most instances, the porosity will be quite small and scattered; most aluminum weldments can possess an appreciable amount of these spherical voids before the joint strength is affected significantly [1706–1708]. All inspection methods, as applicable, can be used to determine internal or surface porosity conditions.

Sources of moisture or hydrocarbons that can produce hydrogen in the weld area are listed below.

1. *Base metals.* lubricants, surface moisture, hydrated oxide films indicated by water stains, etc., must be removed from the joint edges and surfaces of the base metal prior to welding.

2. *Welding rod or electrode.* Surfaces of the filler metal must be free of residual lubricants and hydrated oxide [1709]. The 5XXX series (Al–Mg) filler metals

are most sensitive to forming a hydrated oxide if exposed to moist conditions and should be stored under dry, heated conditions or packaged to prevent exposure.

3. *Welding equipment.* Water-cooled equipment must have controls to shut off the water after welding to avoid a buildup of condensation inside the torch. All connections and hoses in the gas circuit must be leak tight to avoid contamination of the shielding gas, and the inert gas must be a welding grade with tight control of the moisture content.

4. *Atmospheric conditions.* On humid days, moisture in the air can contribute to porosity if the gas shielding pattern is turbulent. Torch nozzle size and shape, torch angle and the gas volume and pressure are important factors in providing proper shielding. When strong drafts or wind gusts are evident, wind-breaks should be employed to avoid disturbance of the shielding pattern.

Weld Inclusions

Metallic inclusions in aluminum welds can be detected by most destructive and nondestructive means. Tungsten, copper and iron are the primary inclusions observed with the gas shielded arc welding processes. Tungsten inclusions most commonly occur as a result of overheating the tungsten or contaminating it with aluminum by dipping it into the molten weld pool or touching it with the filler metal rod. The operator can see when the contamination occurs and should stop welding to clean the tungsten and remove the defective portion of the weld. The tungsten particles do not alloy or fuse with the aluminum weld metal. Thus, these inclusions usually are treated in a similar manner as gas porosity with respect to their effect upon the mechanical performance of the weld.

Copper inclusions occur from melting the end of the contact tube or welding gun nozzle during gas metal arc welding. Melting of the gun nozzle would indicate a need for improved cooling or isolation from the electrical circuitry. Melting of the contact tube can result from improper extension of the tube for a given arc length or from an interruption of the electrode feed. Although the electrode quality (waves or kinks) is often blamed for the meltback, the welding equipment can be a major contributor.

The use of long, straight contact tubes is preferred for aluminum. When used in conjunction with a 'touch-start' or 'slow run-in' wire feed circuitry and a drooping voltampere characteristic power supply to avoid high current surges at arc strikes, arcing in the contact tube and meltbacks can be minimized. A final source of copper contamination with either GTAW or GMAW is the use of a copper backing where openings in the joint edge spacing allow the arc to impinge upon the backing to create melting and alloying of the copper with the aluminum. Removable backing strips of anodized aluminum will provide a good heat sink and avoid the dissimilar metal inclusion in the case of poor joint fitup. Ceramic backings are also useful if a chill effect is not important.

Peripheral cracking can occur during solidification of the weld or under low stress applications in service. The sharp notches or cracks can propagate in service to affect performance of a weldment more significantly than the discontinuities created by gas voids or tungsten inclusions. Surface alloying of copper can also create a corrosion problem when the weldment is exposed to an electrolyte. When a meltback occurs or copper melting is observed, the affected portion of the weld should be removed immediately. This should be a responsibility of the welder, since he knows when the problem occurs; welding normally is terminated at this time to repair the GMAW welding gun.

Ferrous inclusions occasionally are observed in aluminum welds. These result from extraneous sources, such as bristles from a wire brush left in the joint or contamination from a steel backing, such as surface rust. If the steel melts, alloying with aluminum will occur and a variation in density of the radiograph will be apparent at the periphery of the inclusion. Porosity concentrated around the inclusion generally would identify the inclusion as rust, since moisture would be associated with the rust. Ferrous inclusions should be removed from aluminum welds.

Fracture Tests

Nick-break and fillet fracture tests are useful in evaluating the suitability of welding parameters or machine functions at the start of a shift or at periodic intervals during production. These methods are useful particularly when welding equipment is being used by different personnel on different shifts where different machine settings are employed and knowledge of the equipments' performance at the end of the preceding shift is unknown.

The nick-break test is useful in evaluating a groove weld for soundness and can be conducted in a short time at low cost. The specimen is notched with a saw and fractured by hammer blows in a vise or by other suitable means. The exposed surfaces of the fracture can reveal discontinuities such as porosity, incomplete penetration and incomplete fusion. A fillet weld can be made on one side of the joint only and then the joint bent and fractured. The fractured weld metal will reveal porosity, penetration depth and uniformity, and fusion characteristics.

17.1 Problems in Gas Tungsten Arc Welding of Aluminum Alloys

Common problems and their causes are:

Arc-starting difficulty

- Incorrect adjustment of high-frequency spark-gap
- Incomplete welding circuit
- Contaminated tungsten electrode

Inadequate cleaning action by the arc

- Excessive oxide on base metal
- Incorrect adjustment of high-frequency unit or battery bias
- Open-circuit voltage too low

- Inadequate gas shielding caused by:
  Insufficient gas flow
  Spatter on inside of gas nozzle
  Wrong nozzle-to-work distance
  Incorrect position of welding torch
  Drafty environment

Dirty weld bead

- Insufficient shielding gas
- Insufficient gas flow
- Damaged or dirty gas nozzle
- Wrong nozzle-to-work distance
- Incorrect position of welding torch
- Wrong nozzle size (use smallest possible)
- Tungsten electrode not centered in gas nozzle
- Drafty environment
- Impurities in shielding gas because of air or water leakage
- Poor cleaning action by the arc (see causes listed under the section on in adequate cleaning action by the arc)
- Unstable arc
- Electrode contamination
- Dirty workpieces or filler metal

Electrode contamination by aluminum

- Improper filler addition angle or position
- Improper manipulation of torch and filler
- Excessive electrode extension
- Touching of electrode to workpiece

Incorrect electrode contour

- Incorrect electrode size for current
- Incorrect contouring of electrode end before welding
- Wrong electrode material (use pure or zirconiated tungsten electrode with alternating current)

Weld bead contamination by electrode

- Electrode size too small for current
- Improper manipulation of torch
- Aluminum-contaminated electrode
- Wrong electrode material (use pure or zirconiated tungsten electrode with alternating current)
- Insufficient post flow of shielding gas (oxidized tungsten)
- Touching of electrode with filler or to the workpiece

Rough weld bead

- Excessive filler size or nonuniform additions
- Unstable arc
- Improper manipulation of torch
- Incorrect current

Weld bead too wide

- Excessive current
- Welding speed too low
- Arc too long
- Electrode extension too short
- Incorrect position of welding torch

Inadequate penetration

- Wrong edge preparation for the arc characteristics (groove too narrow or shallow)
- Excessive filler metal in weld pool
- Insufficient current
- Arc too long
- Welding speed too high
- Inadequate back gouging

Difficulty in adding filler metal

- Improper feeding angle or position
- Improper manipulation of welding torch or filler metal, or both
- Unstable arc
- Excessive time or too low background current with current pulsation

Poor visibility of arc and weld pool

- Wrong position of work
- Incorrect position of welding torch
- Small or dirty helmet lens
- Wrong size of gas nozzle (use smallest possible)

Overheating of power supply

- Excessive power demand (two similar welding machines can be operated in parallel if the capacity of one is insufficient)
- Poor functioning of cooling fan
- Poor grounding of high-frequency unit
- Poor functioning of by pass capacitor
- Poor functioning of battery bias
- Dirty rectifier stacks (regular maintenance required)

Overheating of welding torch, leads, and cables

- Loose or faulty connections
- Welding torch, leads, or cables too small
- Inadequate cooling-water flow

REFERENCES

1701 Dowd, J. D. 1952. Weld cracking of aluminum alloys. *Welding Journal* 31 (10): 448–s to 456–s.

1702 Dudas, 1. H., and Collins, F. R. 1966. Preventing weld cracks in high strength aluminum alloys. *Welding Journal* 45 (6): 241–s to 249–s.

1703 Scott, M. H., and Gittos, N. F. 1981. Heat-affected zone cracking of Al–Mg–Si alloys. *Welding Journal* 60(6): 95–s to 103–s.

1704 Dowd, J. D. 1956. Inert shielding gases for welding aluminum. *Welding Journal* 35 (4): 207-s to 210–s.

1705 Reicheit, W. R., Evancho, J. W., and Hoy, M. G. 1980. Effects of shielding gas on gas metal arc welding aluminum. *Welding Journal* 59(5): 147–s to 155–s.

1706 Lawrence, F. V, Jr., and Munse, W. H. 1973. Effects of porosity on the tensile properties of 5083 and 6061 aluminum alloy weldments. Welding Research Council Bulletin No. 181.

1707 Lawrence, F. V., Jr., Munse, W.H., and Hurk, J. D. 1975. Effects of porosity on the fatigue properties of 5083 aluminum alloy weldments. Welding Research Council Bulletin No. 206.

1708 Lawrence, F.V., Jr., Munse, W. H., and Hurk, J. D. 1975. Effects of porosity on the fatigue properties of 5083 aluminum alloy weldments. Welding Research Council Bulletin No. 206.

1708 McCarthy, W. A., Jr., lamba, H., and Lawrence, F.V., Jr. 1980. Effects of porosity on the fracture toughness of 5083, 5456 and 6061 aluminum alloy weldments. Welding Research Council Bulletin No. 261.

1709 Martukanitz, R. P., and Michnuk, P. R. 1982. Sources of porosity in gas metal arc welding of aluminum. *Aluminum* 58(5): 276–279.

18

Advantage of Gas Metal Arc Welding

Most of the advantages of GMAW over GTAW stem from the fact that in GMAW direct current electrode positive is used at a high current density. This is possible because the electrode is consumable and is melted during the welding, whereas in GTAW, the current is limited by the melting temperature of the electrode. Heat transfer by the gas metal arc is very efficient.

- **High welding rate**
  Welding speeds two to three times those obtainable by manual GTAW are possible by GMAW, particularly when welding metal more than 10 mm thick.

- **Lower welding cost on metal more than 6 mm thick**

- **Low distortion**
  Because of the high welding speed, which results in rapid chilling of the weld area, distortion using GMAW is generally low. The distortion produced on aluminum due to heat input with GMAW is not more, and is usually less, than that produced on steel of the same thickness when it is welded with flux cored electrodes.

- **Good weld quality**
  The quality of welds produced by GMAW using spray transfer is very high.

- **Good out-of-position welding**
  Because the appreciable arc force projects the weld metal across the arc at a high velocity and is not affected by gravity, welding can be done in any position.

- **High deposition rate**
  High rates of metal deposition are easy to obtain with the large-diameter filler wires (up to 5.5 mm dia.) when they are used with high welding currents.

- **Readily adapted to machine welding**

  Because of its semiautomatic nature, GMAW can be readily adapted to automatic welding for metal from 0.75 mm thick to the thickest commercially available. Automatic GTAW requires good control of joint fit-up, usually within 0.076 to 0.254 mm, depending on material thickness, but GMAW is less sensitive to variations in fit-up.

- **Freedom form radio interference**

  Gas metal arc welding uses direct current, and so it is not necessary to use high-frequency current for arc stabilization. Consequently, there is no radio interference as there may be when using high frequency current with GTAW.

19

Advantages of Gas Tungsten Arc Welding

- **Lower welding costs on metal less than 3 mm thick**
  The equipment used for GTAW is less expensive than that used for GMAW and needs less maintenance.

- **Very thin material can be welded**
  Using a pulsed arc, GMAW can be used to weld metal as thin as 1 mm, but metal as thin as 0.245 mm can be welded by GTAW, provided the workpieces are correctly aligned and held.

- **Excellent weld quality**
  The quality of welds made by GTAW is very high, and the process offers excellent reliability. When welding thick material by GTAW, the filler metal need not be added to the weld until the base metal has been well penetrated; with GMAW, filler metal is added as soon as the arc is struck, which sometimes prevents penetration and causes cold starts.

20

Selection of Process

Generally selection is based on the capability of a process to meet joint requirements and cost. When the thickness of aluminum sheet reaches about 4.5 mm, the use of GTAW with alternating current for single-pass welding becomes prohibitively slow, although joint quality is excellent.

The same joints in this thickness and above can be single-pass welded at a much higher speed using GTAW with DCEN, because of the narrower and deeper penetration obtained, but more careful edge cleaning is needed to ensure high joint quality.

Aluminum sheet 4 mm or more in thickness can also be welded by GMAW using spray transfer and at much higher speeds. *Using DCEP, edge cleaning is less critical than for GTAW.* Joint quality is good and easily meets commercial requirements. Gas tungsten arc welding is capable of meeting the highest standards of nuclear and aerospace applications.

What Industry Wants to Know About Welding Aluminum?

At any discussions of aluminum welding, the same types of questions are asked, year after year, it seems. The following is part of the Questions and Answers from the 1989 Aluminum Welding Seminar in Washington:

Q. If 6061 extrusions are stored outside and exposed to excessive moisture, is drying the material and abrasive cleaning of the weld joint and adjacent area sufficient to ensure weld quality?

A. After drying, a solvent wipe to remove dirt, oil, etc., should be done prior to wire brushing.

Q. Is hot cracking in fillet welds affected by weld profile, i.e. concave vs. convex?

A. Yes, a concave fillet weld would indicate insufficient additions of filler or excessive melting of the base metal or both. This would definitely contribute to hot cracking in the heat-treatable alloys such as 6061.

Q. What do you do when you experience cracking problems when gas metal arc welding of 6061 with 5356 electrode?

A. In order to avoid cracking in restrained joints, about 70% or more of the weld should be 5356 filler metal. To increase the dilution rate of the filler metal, the base metal may have to be beveled more. The degree of the amount of 6061 melted can be reduced via less heat input.

Q.   In arc welding, what are the advantages and disadvantages in using 5356 alloy wire in lieu of 4043 for joining 6061-T6 base metal?

A.   Advantages of 5356 are: higher fillet weld shear strength, higher weld metal ductility, and best colour match after anodizing. Filler alloy 4043 exhibits least sensitivity to weld cracking, is suitable for sustained elevated-temperature service and postweld artificial aging, and provides best corrosion resistance in fresh or seawater immersion.

Q.   How much strength is gained with 5356?

A.   In groove welds thicker than 9.5 mm, 5356 welds provide a 34 MPa higher tensile and compressive yield strength. Minimum shear strength of a fillet weld with 5356 is 117 MPa vs. 79 MPa for 4043. Groove weld metal elongation is 24% for 5356 vs. 16% for 4043 welds in 6061-T6.

Q.   Is the 5356 more subject to cracking than 4043?

A.   Yes. Weld metal should contain a minimum of 50% 4043 filler metal or 70% 5356 to avoid cracking in 6061 base metal.

Q.   When welding 5 × 51 mm angle (one leg only) to 54 mm thick 6061-T6 aluminum with 5356 filler metal, we have experienced cracking or lack of fusion problems on the angle, when 120 °C-plus preheat is used, or on the 50 mm thick base metal, when little or no preheat is used.

A.   Cracking would be expected from excessive melting of the 6061 angle when welding after preheating. Poor fusion to the base plate when not preheating would result normally form an excessive arc length or insufficient amperage. A higher current without preheat should be tried.

Q.   Why doesn't ER4043 filler wire wet and fuse well when welding 6061-T6 aluminum? Gases used have been pure argon and 50% Ar/50% He.

A.   Assuming you are using the GMAW process, poor fusion and wetting would result from: (1) insufficient amperage; (2) excessive voltage or arc length; (3) excessive oxide on the 6061; (4) poor gas shielding or contamination in the gas.

Q.   Is 4043 still regarded as the most suitable filler wire for joining 6061 aluminum to itself?

A.   4043 is used on 6061 for low sensitivity to weld cracking, good corrosion resistance, suitability for sustained elevated temperature service (above 65 °C), and response to re-solution heat treatment for highest strength. The 5XXX series fillers will produce higher as-welded structural strengths and ductility as well as colour match after anodizing.

Q.   What can be done to avoid or reduce spatter during GMA welding?

A.   With all filler alloys, the current must be high enough to produce a spray transfer. With 4043 filler metal, for example, spatter will generally result from too short an arc or too low a voltage. The 5XXX electrode produces large spatter at too short an arc and very fine spatter when the arc is too long. The best arc length produces an occasional crackle with the 5XXX series fillers.

Q. If you had to clean electrodes on a resistance spot weld application for 5052 or 6061 alloy every 10 hits due to pickup of aluminum on the electrode, what might be a basic problem in the process?

A. Excessive electrode pickup can be caused by insufficient cooling of the electrode; rough finish on electrodes; soft temper 5052; thick surface oxides; excessive weld penetration in sheet; insufficient or variable weld force; too small electrode diameter; incorrect electrode materials; and other sources.

REFERENCES

2001 Altshuller, B., *"A guide to GMA welding of aluminum"*, Welding Journal, June 1998, pp. 49–55.

2002 Arc Welding of Aluminum Alloys, Aluminum Association Technical Committee on Welding & Joining.

2003 Fuerschbach, P.W., *"Cathodic cleaning and heat input in variable polarity plasma arc welding of aluminum"*, Welding Journal, Feb. 1998, pp. 76s–85s.

2004 Myers, L.W., "Practical guide to gas flow rates for GTA welding", Welding Journal, Apr. 1997, pp. 75–76.

2005 Welding handbook, Materials and Applications-Part 1, Vol. 3, Eighth Edition, American Welding Society, 1996.

21

Welding Distortion

Weld induced distortion is a familiar problem to all fabricators employing arc welding process. In fact, the weld induced distortion faced by shipbuilders has become a major problem with increased use of aluminum and higher strength steels in thin sections. Distortion problems are often caused at the design stage and they are solved partially during fabrication through experience. The net result is increased man-hours for fairing and rework. On the other hand allowing the distortion to remain also carries with it consequences related to degraded performance, poor fit-up, decrease of the structural integrity and an overall bad appearance. Competitiveness in cost and time can be increased by eliminating or mitigating these distortions during the fabrication process rather than allowing them to accumulate and then removing them. This can be achieved through proper understanding of the distortion mechanism, its type and causes, which will help in developing suitable predictive tool and control mechanisms.

21.1 Distortion Mechanism

Distortion in a welded structure results from the non-uniform expansion and contraction of the weld metal and adjacent base metal during the heating and cooling cycle of the welding process. During heating and cooling in the welding process, thermal strains occur in the weld metal and base-metal near the welded zone. The stresses resulting from these strains produce internal forces, which in turn cause shrinkage, bending, buckling and rotation. The phenomenon of distortion that occurs during a thermal cycle can be explained through the following example. Consider a small cube of steel as shown in Fig. 21.1. As the cube is uniformly heated by some means it expands in all directions as shown in Fig. 21.1 (a). As the heat source is removed and the cube is allowed to cool down, it contracts uniformly to its original dimensions. In the next experiment, let the cube be placed between the jaws of a vice such that it does not fall off (Fig. 21.1(b)). Now if the same cube is again heated as done previously, it will expand only in two directions, and will not be able at expand in the lateral direction because of the restraint provided by the vice. Instead compressive reaction forces will generate as shown in Fig. 21.1(c).

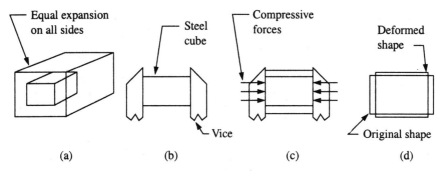

Fig. 21.1   Distortion mechanism

Since volume expansion must occur, the cube expands a greater amount in the vertical direction. Subsequently, on removal of the heating source, the cube cools down and returns to room temperature. Now, in the cooling phase contraction will take place and will occur uniformly from all directions as shown in Fig. 21.1(d). The result is the cube is narrower laterally and thicker in vertical direction. That the cube has become narrower in lateral direction becomes evident from the fact, the cube slips off from the vice. Hence the residual shape of the cube is permanently deformed or distorted. In case of a welded joint, similar expansion and contraction forces act on the weld metal and on the base metal during the heating and cooling phases. The base metal away from the welded zone remains at room temperature through out the welding operation and is not subjected to any expansion or contraction. But this 'cold' part of the base metal acts like a vice holding the welded zone and the adjacent heated base metal and restrains then from free expansion and contraction. As a result as the weld metal solidifies and cools down, it attempts to contract to the volume it would normally occupy at the lower temperature. But because of the restraints from the adjacent cold metal, it can not do so. This leads to stresses of near yield point magnitude in the weld. Under this stress the weld deposit yields and thins out thus adjusting to the volume requirement, of the joint being welded. The stresses, which exceed the yield point, only get relieved by this phenomenon. However when the weld metal and the base metal cools down to the room temperature assuming complete restraint of the base metal the weld deposit tends to have locked-in tensile stresses of near yield point magnitude, which is balanced by compressive stresses in the adjacent base metal or structure as shown in Fig. 21.2. If some of the restraints such as clamps or welded lugs are removed, the locked-in stresses will find partial relief by causing the base metal to deform or distort.

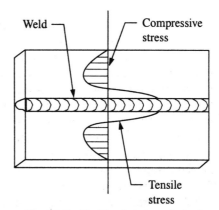

Fig. 21.2   Residual stresses in butt welding

If the tensile weld stresses are not exerted at the neutral axis of the structure, the structure will react by bending (Fig. 21.3) and when this bending occurs in two planes, it results in twisting.

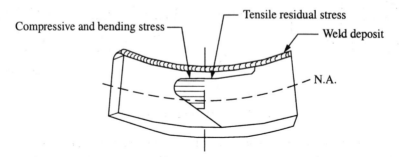

Fig. 21.3 Bending due to weld induced stresses

21.2 Types of Distortion and Their Prediction

Weld induced stresses may cause structural distortions. The typical distortions patterns are shown in Fig. 21.4.

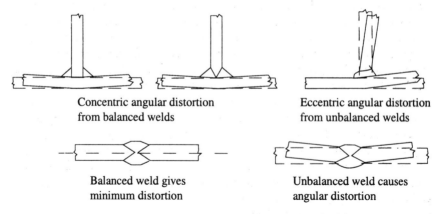

Fig. 21.4 Typical distortions of welded joints

With slow welding speed and with higher number of weld passes for a given weld size, the greater will be the distortion. In view of this, automatic welding with higher arc heat concentration and faster travel speed with higher deposition rate reducing the number of weld passes will reduce the distortion. In case of butt welds square groove angles will cause less distortion because of more even distribution of heat across the plate thickness.

The welding of butts and seams in plate panel assemblies will cause an angular deformation along the weld lines and an overall shrinkage transverse to the weld seams or butts may result. Subsequent welding of stiffeners will produce additional shrinkage and bending stresses in the plate. This will give rise to compressive stresses in the plating perpendicular to the direction of stiffener welding. If these compressive stresses exceed the critical buckling stress of the plate panels, buckling may take place as shown in Fig. 21.4. This phenomenon makes fabrication of thin stiffened panels difficult.

The same distortion mechanism as mentioned earlier, also applies to large assembly units and even to the ship itself. If welding is carried out on top of a long shallow unit, with less rigidity, the unit will deflect upwards at the ends. This explains the phenomenon of the bows and sterns of ships to lift off the keel blocks by several millimeters. On the other hand, if the unit is relatively deep and hence rigid, there will be less tendency to deflect upwards at the ends. This rigidity, in fact can be achieved by a proper erection sequence like, if an upper unit is installed and tack welded to the lower unit and subsequently the welds are made in the lower unit, the top unit will provide added restraint thus further reducing the deflection.

The basic types of distortion that take place during welding are:

- Shrinkage
- Angular deformation
- Rotational distortion
- Buckling distortion

Essentially all the weld induced distortions are caused by the shrinkage force generated due to the thermal loading on the structure. Depending on the pattern of this shrinkage force (compressive forces) various structural deformation like bending, rotation and buckling take place.

21.2.1 Shrinkage

One of the fundamental dimension change that takes place due to welding is in-plane shrinkage of the plates being welded. This shrinkage will take place both in the transverse direction perpendicular to the weld line and in the longitudinal direction parallel to the weld line. Hence the resulting shape of the panel after butt welding will be as shown in Fig. 21.5.

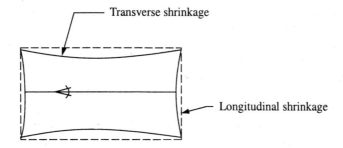

Fig. 21.5 Transverse and longitudinal shrinkage of a butt welded panel

Mechanism

Several investigators have studied the mechanism of shrinkage distortion due to welding and conducted experimental, analytical and numerical analysis of this phenomenon [2101]. The most important finding of these analyses are:

- The major portion of transverse shrinkage of a butt weld is due to contraction of the base plate. Shrinkage of the weld metal itself is only about 10% of the actual shrinkage.

- Transverse shrinkage is generally of much higher order than longitudinal shrinkage.

During welding as the heat from the molten pool gets conducted, the parent metal expands with simultaneous contraction of the weld deposit. This expansion of the parent metal is resisted by the cold part of the base metal not undergoing sufficient temperature change. Thus the plates undergoing welding can be divided into two zones as shown in Fig. 21.6.

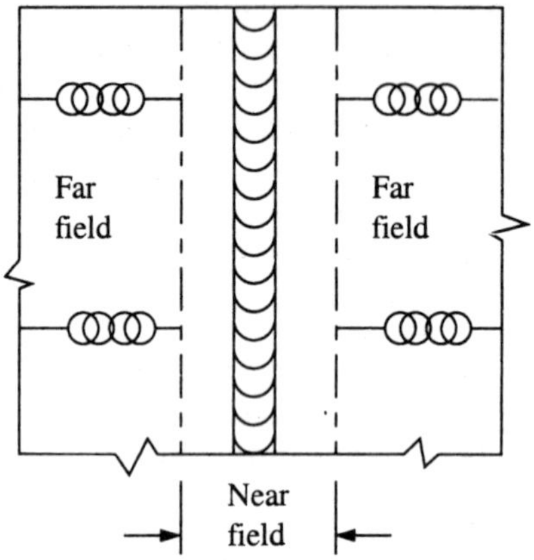

Near Field

Which is directly affected by the thermal cycle and hence experiencing substantial expansion and contraction.

Far Field

Cold part of the plate acting as elastic restraints and not directly affected by the thermal cycle and hence does not undergo a direct expansion or contraction.

Fig. 21.6   Near- and Far-field zones of the plates undergoing welding

Mechanism of transverse shrinkage can be explained as follows:

- The base metal in the nearfield zone expands during the heating cycle.
- The expansion is constrained by the cold part of the base metal i.e. farfield zone.
- Compressive forces are generated. This can be modeled as the expansion forces exerted by the nearfield is stored as energy in a set of springs along the boundary of the nearfield.
- During the subsequent cooling cycle, contraction of the nearfield takes place and simultaneously the energy stored in the said springs gets released thus adds to the contractions forces.
- Under the action of these forces, the base metal along with the weld deposit with low yield point because of elevated temperature yields and plastic flow of material takes place giving rise to transverse shrinkage and at the same time thickening of the base metal near the weld deposit.

Estimation of Transverse Shrinkage

Rigorous theoretical determination of thermal distortion is complicated by the creation of localised transient zones of plasticity as the heat source moves along the workpiece. Further difficulties arise in accounting for heat losses due to convection and radiation from the heated areas, difficulty of defining realistic boundary conditions as well as other arbitrary and variable effects such as weld penetration. These factors add a random element to any method of predicting thermal stresses. It is therefore unlikely that any improvement would accrue from more complex or rigorous analysis without disproportionate trouble and cost. The physical basis of thermal distortion and the factors affecting it are not difficult to understand, and therefore empirically determined formulae are adequate for practical purposes.

Transverse Shrinkage

This is important where unrestrained welded panels have to be fitted accurately to mating assemblies. If shrinkage is restrained, tensile stresses are generated perpendicularly to the weld direction. Spraragen [2102–2103] has proposed a very simple empirical formula for the transverse shrinkage of a butt weld:

$$S = 0.2\, A_w/t_h + 0.05r \qquad (21.1)$$

where A_w is the cross-sectional area of the weld and r the initial root gap.

Measurements of the free transverse shrinkage by White [2103–2104] have led to the following empirical formula for a single pass butt weld:

$$S = \frac{3\eta\left(\dfrac{Q}{S_w}\right)\left(1 - \dfrac{z}{2t_h} + 0.2r\right)}{t_h}\, mm \qquad (21.2)$$

where r is the root gap in mm and z the unfused depth of root (Fig. 21.7).

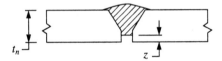

Fig. 21.7   Unfused depth of root

The values of z depends on the penetration which if not known can be calculated from Leggatt's formula [2104–2105]

$$z = t_h - \left[\frac{0.3I^4}{V^2 S_w}\right]^{1/3} \qquad (21.3)$$

If z and r are not known, a simple formula may be used as

$$S = \frac{4.4\eta(Q/S_w)}{t_h} \qquad (21.4)$$

or if the penetration is not complete, then

$$S = \frac{4.4\eta(Q/S_w)}{t_h} - 0.175 \qquad (21.5)$$

These formulae corroborate Capel's study [2105–2106] which suggested

$$S = \frac{2.9(Q/S_w)}{t_h} \qquad (21.6)$$

The shrinkage across *T*-fillet and cruciform weld joints is also given in [2103–2104],

$$S = \frac{3\eta(Q/S_w)nc}{2t_h + t_w} \qquad (21.7)$$

where n = total number of weld passes in joint
t_w = thickness of web
$c = 2$ for shrinkage of flange (plate)
$c = 1$ for total shrinkage across joint in web direction.

For multi-pass welds the transverse shrinkage may be estimated as the cumulative sum of that due to each individual pass. In the case of a butt weld, there is a slight complication due to angular contraction known as "wrap-up". The nth pass produces a shrinkage according to equation (21.5) of

$$S_n = 3.5 \left(\frac{Q}{S_w t_n} - 0.05 \right) mm$$

where t_n is the thickness after the nth pass.

This occurs at the current mid-weld $\frac{1}{2} t_n$; the shortening at the centre of the plate is

$$S = \Sigma S_n + \frac{1}{2} \phi_n(t_h - t_n) \qquad (21.8)$$

where ϕ_n is the angular wrap-up discussed below.
A simpler thumb rule is

$$S = 1 + 0.044t_h \qquad (21.9)$$

where S and t_h are in mm.

21.2.2 Out-of-Plane Distortion
The major distortion problem in the fabrication of stiffened panels is related to that of out-of-plane distortion caused by angular changes along the fillet welds (Fig. 21.8).

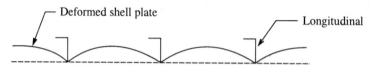

Fig. 21.8 Angular deformation due to fillet welding of stiffeners

At the same time, the initial distortions and residual stresses adversely affect the buckling strength of the stiffened panels. The corrugations often observed in bottom shell or side shells are generally caused by buckling failure of the stiffened panels (Fig. 21.9).

Fig. 21.9   Buckling deformation due to fillet of stiffeners

21.2.2.1   Angular Distortion

Angular distortion due to butt welding of two plates takes place as a result of a bending moment caused by the formation of non-uniform shrinkage forces across the plate thickness in the regions near the weld line as shown in (Fig. 21.10).

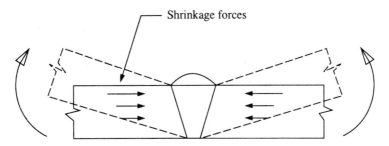

Fig. 21.10   Angular deformation due to butt welding

The angular change in a butt weld depends on the joint geometry. A square butt leading to a more uniform heat distribution in thinner plates yields minimum angular deformation, whereas single V groove welds lead to a higher level of angular distortion. However in double V groove welding the resulting angular deformation is minimum. Severe angular distortion is caused by fillet welds. In case of plates with no external constraints, the angular distortion due to fillet welds is shown in Fig. 21.11.

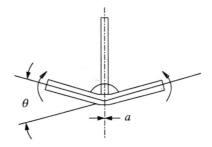

Fig. 21.11   Angular distortion in a free fillet weld

The bending of the plate is caused by a moment generated due to the nonuniform distribution of shrinkage forces resulting in nonuniform plastic deformation in the thickness direction produced during welding in the near vicinity of the deposited metal. The plastic deformation being localized in small areas near the weld, the rest of the plate away from the weld remain virtually straight. Therefore, the angular distortion of a free joint can be characterized by angular change ϕ as shown in Fig. 21.11.

When several members are fillet welded with no external constraints as in the case of a flat narrow plate panel stiffened by longitudinals the plate bends at each joint and forms a sort of polygon, (Fig. 21.12).

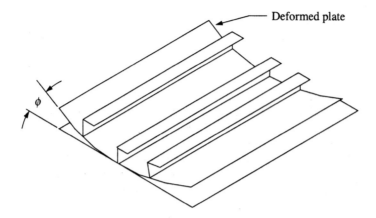

Fig. 21.12   Angular distortion in an unrestrained stiffened panel

On the other hand, if the above referred stiffened panel has some constraints e.g. deep transverse members, then a different type of distortion will result (Fig. 21.13).

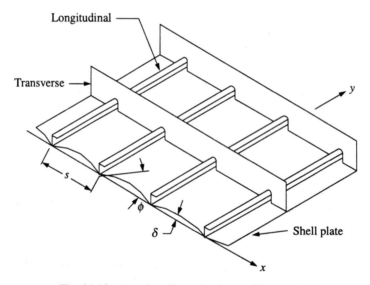

Fig. 21.13   Angular distortion in a stiffened panel

In a situation where the sizes of all the welds of the longitudinals are same, the distortion in all the spans will be equal and the distortion δ can be expressed as [2101]

$$\delta/s = \left[\frac{1}{4} - \left(\frac{x}{s} - \frac{1}{2} \right)^2 \right] \phi \qquad (21.11)$$

Therefore, maximum distortion at the middle of two longitudinals will be given by

$$\delta_0 = \frac{1}{4}\phi s \qquad (21.12)$$

The angular changes ϕ in case of a restrained panel (Fig. 21.13) will be smaller than that of an unrestrained panel (Fig. 21.12). The amount ϕ will also depend on the plate thickness t_h and the span s. It was found that the following relationship exists:

$$\phi = \frac{\phi_0}{1 + \left(\dfrac{2D}{s}\right)\left(\dfrac{1}{c}\right)} \qquad (21.13)$$

where $D = \dfrac{Et_h^3}{12(1 - v^2)}$ c = coefficient of rigidity for angular changes is a parameter that depends on welding conditions and plate thickness. The values of c for aluminum are given in Table 21.1 [2101]. To convert from w to the fillet leg length a (mm), the following formula may be used

$$w = (a^2/2) \times 10^{-2} \times \rho/\eta_d \qquad (21.14)$$

where w = weight of electrode consumed per weld length (g/cm)
$\quad$ a = leg length (mm)
$\quad$ ρ = density of weld metal (g/cm^3)
$\quad$ η_d = deposition efficiency

It has been observed that angular distortion is maximum for plates of 8-10 mm thickness whereas it is much less for thinner plates as well as for thicker plates. This is because for thinner plates, they are heated more evenly in the thickness direction, thus reducing bending moment. When the plates are thicker than 10 mm, the angular distortion reduces because of the increase of rigidity.

Deflections Due to Welding Distortions

There are two distinct causes of deflections in welded plating. One is the residual compression which increases the likelihood of local plate or stiffener buckling. In thin plating such buckling may be evident even before load is applied to the structure. The second cause of deflections is angular deformation at the welds and the effect can be estimated by assuming that the plating dishes into a parabolic shape.

A plate of width b experiencing an angular deformation of $\frac{1}{2}\phi$ at each edge will thus have a central deflection of about

$$w_{max} = 0.125b\phi/57.3 \qquad (21.15)$$

and a radius of curvature

$$R = \frac{b^2}{8w_{max}} = \frac{57.3b}{\phi} \qquad (21.16)$$

Table 21.1 Values of angular rigidity coefficient for aluminum

Amount of weld

Fillet leg length a (mm)	Weight of consumed electrode per unit weld length w (g/cm)	$\log_{10} w$	$t_h = 3.18$ (mm)	$t_h = 6.4$ (mm)	$t_h = 9.5$ (mm)	$t_h = 12.7$ (mm)	$t_h = 15.9$ (mm)	$t_h = 19.1$ (mm)
8.969	1.122	0.05	57	782	14,390	22,800	31,000	78,400
9.567	1.259	0.1	55	762	13,600	20,800	25,300	72,500
10.660	1.585	0.2	52	725	7,900	17,000	18,000	31,800
11.960	1.995	0.3	49	686	5,600	13,800	12,900	22,200
13.420	2.512	0.4	46	645	4,300	11,000	9,200	17,000
15.057	3.162	0.5	43	608	3,600	8,900	6,9900	13,500

$w = (a^2/2) \times 10^{-2} \times 2.65/0.95 = 0.0139 \, a^2.$

Deflections Due to Residual Stress

The compressive stresses which equilibrate the tendon force effectively degrade the buckling strength and the reduction in the elastic critical stress $\Delta\sigma_{cr}$ for a rectangular plate has been shown [2108–2109] to be,

$$\Delta\sigma_{cr} = \sigma_y \sin\left(\frac{\sigma_r}{\sigma_y}\right) \tag{21.17}$$

where σ_y is the yield stress and σ_r the compressive residual stress.

For thin plates, it is apparent that the reduction in the critical stress might well exceed the original critical stress. When this happens the plate starts to buckle before any load is applied externally, quite apart from any distortion directly due to weld shrinkage. The conditions under which this occurs can be calculated by combining equations (22.15) and (22.17) and relating equation (21.17) to the initial buckling stress:

$$\sigma_r = \frac{200000\eta\ Q/S_w}{bt_h}\ N/mm^2 \tag{21.18}$$

Therefore

$$\Delta\sigma_{cr} = \sigma_y \sin\frac{200000\eta\ Q/S_w}{\sigma_y bt_h}$$

but

$$\sigma_{cr} = \frac{4\pi^2 Et_h^2}{12(1-v^2)b^2}$$

Equating this to $\Delta\sigma_{cr}$, the net welding heat input which will just cause buckling distortion in aluminum plating is

$$\eta Q/S_w = 0.005\ bt_h\sigma_y\ \text{arc}\sin\left(\frac{4\pi^2 Et_h^2}{12(1-v^2)b^2\sigma_y}\right)\ kJ/m \tag{21.19}$$

If the compressive residual stress σ_r is less than about a third of the yield stress (very likely in practice), equation (21.19) reduces to (for steel)

$$\eta Q/S_w = \frac{3.7t_h^3}{b}\ kJ/mm \tag{21.20}$$

where b and t are both in mm.

If this is related to equation (22.18) or (22.19), the likelihood of buckling due to residual stress can be readily established.

Example

Consider a 5 mm thick plate, stiffened longitudinally at 400 mm intervals by fillet welded T sections. Assuming a MMW process with arc efficiency of 0.7 and the fillet welds are performed consecutively, find the maximum permissible fillet size before onset of buckling due to residual stress.

From the following equation, the maximum net heat input is

$$\eta Q/S_w = \frac{3.7 \times 5^3}{400} = 1.16\ kJ/mm$$

Since the fillet welds on each side of the stiffener are performed consecutively, the heat input from the first weld may be assumed to be washed out, and for a MMW process taking $C = 17$ from equation (22.18) one obtains

$$a(a + 1) = .7 \times 1.16 \times 17 = 28.17$$

i.e. $a = 4.8$ mm

Therefore 4.8 mm is the maximum fillet size, beyond which buckling is expected due to residual stress.

The conclusion is in agreement with shipyard practice where this type of distortion is frequently observed on thin stiffened plates but the magnitudes of the distortions are often surprisingly large even when the welds are properly designed. This phenomenon can be explained by the influence of initial stresses in the elements to be welded [2109].

The influence of initial stress upon the postwelding distortion can be expressed by

$$m = 1 - \frac{\ln(1 + \beta)}{\ln 2}, \qquad (21.21)$$

where $\beta = \dfrac{\sigma_0}{\sigma_y}$, σ_0 is the initial stress and σ_y the yield stress.

As may be seen from Fig. 21.14 for $\sigma_0 = 0$, $m = 1$; for $\beta = 1$ (initial loading equal to yield), $m = 0$, i.e. no distortion at all after welding but for $\beta = -0.5$, $m = 2$, i.e. twice the distortion.

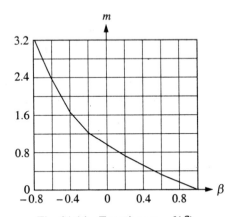

Fig. 21.14   Function $m = f(\beta)$

In the case of slender stiffened plates which buckle due to welding, the influence of initial loading is much more significant. Small tensile stresses not only can diminish distortion but above all can prevent buckling. On the other hand, even small compressive stress causes not only substantial growth of distortion but also promotes buckling. The influence of initial loading upon postwelding distortions has been confirmed experimentally [2109] as well as experienced in shipyard practice. Compressive stresses can exist in plates to be welded. Both tensile and compressive stresses form at the end of hot rolling process. These stresses do not disappear unless mangling is properly performed before further processing them. These stresses may also form when deflected elements are forcibly tacked together for alignment to provide a proper welding gap.

21.2.2.2 Buckling Distortion

Use of aluminum and thin section steels have made the required levels of fairness in fabricated structures more difficult to achieve. Competitiveness in cost and time can be increased by eliminating or mitigating these distortions during the fabrication process rather than allowing them to accumulate and then removing them. The major part of out-of-plane distortion in fabrication of stiffened panels is due to buckling. At the weld region the residual stresses caused by welding are tensile and of yield magnitude. Self equilibrating compressive residual stresses develop in the rest part of the plate away from the tensile zone. When the compressive stresses on the panel exceed the critical buckling stress of the panel, buckling occurs.

The severe temperature gradients developed during welding produce thermo-mechanical coupling as a consequence of thermal expansion and shrinkage. At locations away from the weld zone, the high stress levels at the weld zone produce plastic deformations. These plastic deformations lead to residual stresses and buckling upon cooling. In studying weld induced out-of-plane distortions in thin-panel structures, it is important to first determine whether the distortion is a result of angular deformation or produced by buckling. Buckling distortion differs from angular distortion in that:

- There is more than one stable deformed shape.
- The magnitude of distortion in buckling is much greater.

Since the amount of buckling distortion is large, the best way to avoid it is to properly select such structural parameters as plate thickness, stiffener spacing and welding parameters. Experiments conducted at Kawasaki Heavy Industries indicate the heat input-deflection relationship is similar to the load deflection relationship in a buckling experiment.

It is interesting to note that the experimental results indicate the existence of a critical buckling heat input for a given test condition. The critical heat input decreases as plate thickness decreases and the free span increases.

Simplified close-form analytical solutions have been proposed for the predication of critical buckling load [2101]. These solutions involve an idealization of the residual stress state on a flat plate for various boundary conditions and weld locations. However these solutions have severe limitations for application in shipbuilding situations because of the absence of stiffeners in the formulation of the critical loads.

Finite-element method has been used in an attempt to predict distortions and residual stresses due to welding. The simulation of welding process involves performing a thermo-mechanical analysis. Transient nonlinear thermal analysis and small deformation quasi-static elasto-plastic analyses have been reported by a number of investigators [2110–2117]. However the majority of these simulations focus on simple structural elements in the area adjacent to the weld. Brown and Song [2118, 2119] have shown that the interaction between the weld zone and the structure can have a very significant effect on the accumulated distortion. In many cases, the overall structural arrangement influence the state of distortion and stress, which is much different from the one predicted by analysis of the weld zone alone. In addition to these some distortion modes, including buckling, cannot be represented by two-dimensional models. On the other hand, the use of fully three-dimensional models to perform thermo-mechanical weld simulations of complex structures is computationally prohibitive and impractical [2120].

Michaleris et al. in [2121] has suggested an efficient and effective numerical technique for predicting weld-induced buckling distortion. The technique combines three-dimensional structural analysis, with two-dimensional welding simulations. Small-deformation, eigen-value, and large-deformation analyses are performed progressively using a thermal load applied to the welded region. The welding conditions and joint configuration are the input to the two-dimensional nonlinear, transient thermal analysis, where the heat flow is simulated in the plane perpendicular to the welding direction. The output from this analysis is the temperature distribution. In the mechanical analysis, which follows the thermal analysis, the elastic-plastic deformations are analysed. The distortion and residual stresses are the output of the mechanical analysis. The severe temperature gradients developed during welding produce thermo-mechanical coupling as a consequence of thermal expansion and shrinkage. At plate regions away from the weld zone (far field) the stress state remains elastic. However, the high stress levels at the weld zone produce plastic deformations. These plastic deformations lead to formation of residual stresses and distortion upon cooling. Eigenvalue analysis can be performed to determine the critical buckling load and the buckled mode shape. However, in order to predict the distortion magnitude a large deformation static analysis is required.

REFERENCES

2101   K. Masubuchi, *Analysis of Welded Structures,* Pergamon Press, 1980.
2102   Mandal, N.R. and Sundar, C.V.N., "Analysis of Welding Shrinkage", *Welding Journal,* June (1997).
2103   Spraragen, W. and Ettinger, W.G., "Shrinkage distortion in welding", *Welding Journal,* 29, nos. 6 and 7, Research Supplement, 292s–294s and 323s–335s (1950).
2104   White, J.D. and Dwight, J.B., "Residual stresses in large stiffened tubular structures", *CUED Report,* C-Struct/TR67, 1978.
2105   Leggatt, R.H. and White, J.D., "Predicting shrinkage and distortion in a welded plate", *Conference on Residual Stresses, Welding Institute,* 1977, Paper 16, pp. 119–132.
2106   Capel, L., "Aluminum welding practice", *British Welding Journal,* Vol. 8 No. 5, 1961, pp. 245–248.

2107 White, J.D., et al., "Weld shrinkage prediction", *2nd Boss Conference*, 1979, pp. 245–266.

2108 Ractliffe, M.A., "The basis and essentials of thermal residual distortion in steel structures", *RINA*, 1983.

2109 Dwight, J.B. and Ractliffe, A.T., "The strength of thin plate in compression", Paper No. 1 from *Thin Walled Steel Structures*, Crosby-Lockwood, 1969, pp. 61–66.

2110 Borzecki, T. and Woloszyn, A., "Experimental tests on the influence of external loading on the reduction of post-welding distortions of thin-walled structures", *Report No. 26, Ship Research Institute,* Technical University of Gdansk, Poland.

2111 Argyris, J.H., Szimmat, J., and Wiliam, K.J., "Computational Aspects of Welding Stress Analysis", *Computer Methods in Applied Mechanics and Engineering,* Vol. 33, 1982, pp. 635–666.

2112 Bibby, M.J., Goldak, J.A., Szyskowicz, M.S., and Jefferson, I., "Computational developments in the Watt Transient Microstructure Algorithm", *Modeling of Casting and Welding Processes,* A.F. Giamei and G.J. Abdaschian, Eds., IV, Feb. 1988, pp. 81–91.

2113 Chakravarti, A. Malik, L.M., and Goldak, J., "Prediction of Distortion and residual Stresses in Panel Welds", *Computer Modeling of Fabrication Processes and Constitutive Behavior of Metals,* Ottawa, 1986. pp. 547–561.

2114 Free, A.J. and Goff, R.F.D., "Predicting Residual Stresses in Multipass Weldments with the Finite element Method," *Computers and Structures,* Vol. 32(2), 1989, pp. 365–378.

2215 Michaleris, P., Tortorelli, D.A., and Vidal, C.A. "Analysis and optimization of Weakly Coupled Thermo-Elasto-Plastic system with Applications to Weldment Design," *International Journal for Numerical Method in Engineering*, Vol. 28, 1995, pp. 1259–1285.

2116 Papazoglou, V.J. and Masubuchi, K., "Numerical analysis of Thermal Stresses During Welding Including Phase transformation Effects", *Journal of Pressure Vessel Technology,* Transactions of the ASME, Vol. 104, Aug. 1982, pp. 198–203.

2117 Tekriwal, P. and Mazummder, J., "Transient and Residual Theral Strain Stress Analysis of GMAW", *Journal of Engineering Materials and Technology,* Transactions of the ASME, Vol. 113, July 1991, pp. 336–343.

2118 Ueda, Y., Muralawa, H., Gu, S. Mm., Okumoto, Y., and Kammichika, R., "Simulation of Welding Deformation for Precisio Ship Assembling (Report 1), In-plane Deformation of Butt welded Plate," *Transactions of JWRI*, Vol. 21(2), 1992, pp. 265–275.

2119 Brown, S.B. and Song, H., "Finite Elements Simulation of Welding of Large Structures", *Journal of Engineering for Industry,* Vol. 114, Nov. 1992(a).

2120 Brown, S.B. and Song, H., "Implications of Three-Dimensional Numerical Simulations of Welding of Large Structures", Welding Research Supplement, Welding Journal, Vol. 71(2), Feb. 1992(b), pp. 55s–62s.

2121 Debiccari, A., Michaleris, P., "A predictive Technique for Buckling Analysis of Thin Section Panels due to Welding", *Journal of Ship Production,* Nov. 1996.

Furthermore, residual stresses can also be classified, according to the mechanism of formation, to those produced by structural mismatching and those produced by an uneven distribution of non-elastic strains (including plastic and thermal strains).

In case of welding, residual stresses can be classified as being macroscopic in scale and as being produced by uneven distributions of non-elastic strains. Finally, it should be noted that since residual stresses exist without any external loads they should always satisfy force and moment equilibrium, i.e. they should be self-equilibrating.

22.2   Welding Residual Stress Formation

To physically understand how residual stresses are formed during welding, the simple case of a straight bead-on-plate weld will be described in some detail [2201]. Fig. 22.2 shows schematically the changes of temperature and stresses that occur during such a process. The welding arc, which is moving at a speed S_w, is presently located at the origin O (Fig. 22.2(a)). Fig. 22.2(b) shows the temperature distribution along several cross sections. Along section A–A, which is ahead of the welding arc, the temperature change due to welding, ΔT, is almost zero. Along section B–B, which crosses the welding arc, the temperature change is extremely rapid and the distribution is very uneven. Along section C-C, which is some distance behind the welding arc, the distribution of temperature change is as shown in Fig. 22.2(b)-3. Along section D-D, which is very far from the welding arc, the temperature change due to welding again diminishes.

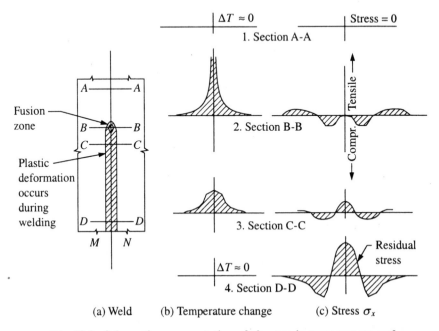

Fig. 22.2   **Schematic representation of changes in temperature and stresses during welding**

Figure 22.2(c) shows the distribution of stresses along these sections in the x-direction, σ_x. Stress in the y direction, σ_y, and shearing stress, τ_{xy}, also exist in a two-dimensional stress field. Along section *A-A*, thermal stresses due to welding are almost zero. The stress distribution along section *B-B* is shown in Fig. 22.2(c)-2. Because molten metal will not support a load, stress underneath the welding arc is close to zero. Stresses in regions, a short distance from the arc are compressive, because the expansion of these areas is restrained by the surrounding metal where the temperatures are lower.

Since the temperatures of these areas are high and the yield strength of the material low, stresses in these areas are as high as the yield strength of the material at corresponding temperatures. The magnitude of compressive stress passes through a maximum with increasing distance from the weld or with decreasing temperature. However, stresses in areas away from the weld are tensile and balance with compressive stresses in areas near the weld. In other words

$$\int \sigma_x dy = 0$$

across section *B-B*. This equation neglects the effect of σ_y and τ_{xy} on the equilibrium condition. Thus, the stress distribution along section *B-B* is as shown in Fig 2.2(c)-2. Stresses are distributed along section *C-C* as shown in Fig. 22.2(c)-3. Since the weld metal and base metal regions near the weld have cooled, they contract and cause tensile stresses in regions close to the weld. As the distance from the weld increases, the stresses first change to compressive and then become tensile.

Fig. 22.2(c)-4 shows the stress distribution along section *D-D*. High tensile stresses are produced in regions near the weld, while compressive stresses are produced in regions away from the weld. This is the usual distribution of residual stresses that remain after welding is completed. The cross-hatched area, *M-N* in Fig. 22.2(a) shows the region where plastic deformation occurs during the welding thermal cycle. The egg-shaped region near the origin O indicates the region where the metal is melted. The region outside the cross-hatched area remains elastic during the entire welding thermal cycle.

22.3   Sources of Residual Stresses

The difference in shrinkage of differently heated and cooled areas of a welded joint was identified as the primary cause of residual stress formation, resulting in high longitudinal stresses, σ_x, in the weld metal. Similar tensile stresses, σ_y, arise in the transverse direction, too, but of smaller magnitude.

One source is the uneven cooling in the thickness direction of the weld. Surface layers of the weld and the highly heated areas close to it usually cool more rapidly than the interior. Especially in the case of thick plates. Thermal stresses thus arise over a cross section which can lead to heterogeneous plastic deformation and hence to residual stresses. These quenching residual stresses are expected to be compressive at the surface of the highly heated areas and to self-equilibrate with the tensile ones in the inner regions.

The other source of residual stresses comes from the phase transformations that occur during cooling; in the case of steel, for example, austenite is transformed into ferrite, bainite, or martensite, or a combination of them. These transformations are accompanied by an increase in specific volume causing the material being transformed (in the weld metal and the heat affected zone) to tend to expand. This expansion, however, is hindered by the cooler material which is not being transformed, inducing compressive stresses in the other regions. Thermal gradients thus generate differential strains which if high enough can cause yielding and result in residual stress. The total residual stresses due to welding can thus be found by combining the effects of the aforementioned three sources.

22.4   Methods for Predicting Residual Stresses

The incompatible non-elastic strains produced in the weldment as a result of the non-uniform temperature distribution are formed in a very complex manner as discussed in previous sections. Simulating the phenomena that occur during the welding process can make prediction of the residual stress distribution.

To accurately simulate the thermo-mechanical behavior of a weld, or for this matter any phenomenon that involves both thermal and stress analysis, one should start from first principles, which in this case is the first law of thermodynamics. This would mean that one would have to solve problem containing mechanical and thermal coupling, a fact that makes the analysis extremely complicated, if not impossible, on the basis of the present state of knowledge, expect in cases where thermoelastic modeling is sufficient. Initial attempts at formulating and investigating the more complex problem of coupled thermoplasticity have been undertaken by Mroz [2202], but the whole subject area is still developing. It becomes necessary, therefore, to uncouple the thermal and mechanical parts of the welding problem and solve each one separately. The assumptions required for this uncoupling have been examined by Hibbitt and Marcal [2203]. The most critical ones are the neglect of dimensional changes and the neglect of cross-coupling between thermal and mechanical work.

Based on these assumptions the problem can be solved in two steps. First, the heat flow during welding is analyzed. The obtained temporal and spatial temperature distributions are then used as one of the inputs for the subsequent strain and stress analysis. Various analytical and numerical methods that can be used in each step are detailed in the next sections.

Problem Characteristics

A typical welding configuration is shown in Fig. 22.3. A welding arc is travelling with a speed S_w between two plates, causing them to coalesce by providing filler metal.

The four parts of a weld that are subjected to different thermal histories are shown in Fig. 22.4. Part 1 constitutes the filler metal which is deposited molten and later solidifies as cooling begins. Part 2 is the part of the joint that metals and later resolidifies during cooling. Both these parts define the fusion zone. Part 3 is the heat-affected zone (HAZ), defined for steel as that part of the joint

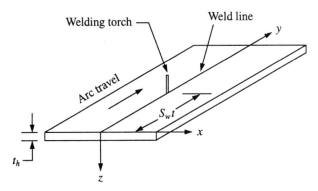

Fig. 22.3 A typical welding configuration

in which the maximum temperature reached is above the A_1 temperature but below the solidus temperature. A_1 temperature is defined for steel as that temperature at which the eutectoid reaction takes place. Under equilibrium conditions this temperature is equal to 723°C. Solidus is defined as that temperature above which the liquid phase is also present. Finally, part 4 is the base metal. These four thermal histories and the related changes in physical properties are shown in Fig. 22.5 for the case of a single pass weld [2204].

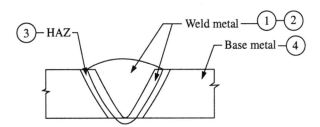

Fig. 22.4 Four parts of a weld subjected to different thermal histories

22.4.1 Analysis of heat transfer during welding

The importance of accurately predicting the temperature distribution during welding has been recognized for many years by both scientists and engineers working with welding problems. This importance stems from the fact that most of the phenomena subsequently encountered, such as residual stresses, distortion, and metallurgical changes, have their origin in the uneven temperature distribution and the fast heating and cooling rates that occur during the welding operation.

Early attempts at solving the problem of heat flow during welding were analytical in nature. As a consequence several simplifying assumptions had to be made to allow for the solution of the highly nonlinear governing partial differential equation and the accompanying boundary conditions. As with any complex engineering problem, the choice of whether to use an analytical or a numerical solution has to be based on a cost versus accuracy tradeoff. Analytical solutions are much more inexpensive and, though not as accurate as the numerical ones, provide nevertheless for the establishment of the general laws and thus facilitate

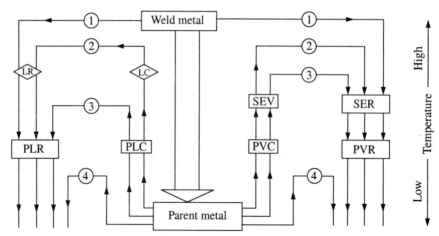

Fig. 22.5   Thermal histories and related changes of physical properties

$\langle LC \rangle$ — Absorption of latent heat during melting

$\langle LR \rangle$ — Release of latent heat during solidification

PLC — Absorption of latent heat during phase transformation
PLR — Release of latent heat during phase transformation
PVC — Volume shrinkage during phase transformation
PVR — Volume expansion during phase transformation
SEV — Yield stress and Young's modulus vanish
SER — Yield stress and Young's modulus recover

a good understanding of the phenomena involved. On the other hand, the more expensive numerical solutions are necessary whenever accuracy is of paramount importance, as for example when a metallurgical characterization of the weld metal and the HAZ is needed, or when a subsequent stress analysis to determine the transient strains and residual stresses is required.

22.4.1.1   Closed Form Analytical Solutions

The first exact analytical solutions of the problem of heat flow during welding were obtained by Rosenthal [2205–2206] in the late 1930's and early 1940's. Rosenthal solved the conventional heat conduction differential equation for constant point, line, and plane heat sources moving at a constant speed with respect to a fixed Cartesian coordinate system. To facilitate easier handling of the problem, he assumed that welding was performed over a sufficient length so that the temperature distribution around the heat source would not change if viewed from a coordinate system moving with the heat source. This phenomenon is called quasi-stationary or quasi-steady state. Additional assumptions made are as follows:

- The physical properties of the conducting medium are constant.
- The heat losses through the surface of the conducting medium to the surrounding atmosphere are neglected.

- Heat generated in electric welding by the Joule effect is negligible.
- The phase changes and the accompanying absorption or release of latent heat in the conducting medium are neglected.
- The conducting medium is infinitely large in the two dimensional case (line heat source) and semi-infinitely large in the three-dimensional case (point source).

Based on the foregoing assumptions Rosenthal developed the following exact solution for the two- and three-dimensional cases, respectively

$$T - T_0 = \frac{Q}{2\pi k t_h} \cdot e^{-\lambda S_w \xi} \cdot K_0(\lambda S_w r) \tag{22.1}$$

$$T - T_0 = \frac{Q}{2\pi k} \cdot e^{-\lambda S_w \xi} \cdot \frac{e^{-\lambda S_w R}}{R} \tag{22.2}$$

where

T_0 = initial temperature; Q = total heat input
t_h = plate thickness; $\xi = x - S_w t$
S_w = welding speed; t = time
(x, y, z) = fixed cartesian coordinate system
$r = (\xi^2 + y^2)^{1/2}$; $R = (\xi^2 + y^2 + z^2)^{1/2}$
k = thermal conductivity
$(1/2) \lambda = k/\rho c$ = thermal diffusivity
ρ = density; C_p = specific heat
$K_0(x)$ = modified bessel function of second kind and zero order

For the case of thin plates (two-dimensional solution) Rosenthal [2205] and other investigators have suggested that heat losses through the surface to the surrounding atmosphere might have to be taken into account by replacing in the Bessel function of equation (22.1), the factor λS_w by

$$\left[(\lambda S_w)^2 + \frac{h_1 + h_2}{k t_h} \right]^{1/2} \tag{22.3}$$

where h_1 and h_2 are the heat-transfer coefficients, assumed constant, at the top and bottom of the plate respectively.

Furthermore, by using the so-called method of images or "image source method" [2205], one can get solutions for the cases of large but finite thickness and/or finite breadth plates. For example, a three-dimensional solution for laying a weld bead on the top of a finite-thickness plate with adiabatic boundary conditions can be obtained by adding an infinite series to equation (22.2), yielding

$$T - T_0 = \frac{Q}{2\pi k} \cdot e^{-\lambda S_w \xi} \left\{ \frac{e^{-\lambda S_w R}}{R} + \sum_{n=1}^{x} \left[\frac{e^{-\lambda S_w R_n}}{R_n} + \frac{e^{-\lambda S_w R_n'}}{R_n'} \right] \right\} \tag{22.4}$$

where $R_n = [\xi^2 + y^2 + (2nt_h - z)^2]^{1/2}$

$R_n' = [\xi^2 + y^2 + (2nt_h + z)^2]^{1/2}$

Following these initial development many investigators tested the validity of the proposed equations experimentally. A through exposition of these works can be found in Myers et al. [2207]. Looking at data on thermal conductivity, specific heat, and density one finds that all these parameters are highly dependent upon temperature, thus making the constant-properties assumption of the analytical solutions unrealistic, especially for the regions close to the heat source where the material exhibits very high temperatures. It il»desirable to predict the high-temperature region as accurately as possible, since it is this region that is directly related to the size of the plastic zone and the accompanying residual stresses and distortion.

The iteration method is adopted to take into account the temperature dependence of material properties. The fundamental heat source solution with material properties at some temperature, say 200°C provides the first approximate solution at a particular point. This temperature is compared with the initial guess and if the two temperatures disagree by more than 0.5°C, new properties are found for a temperature halfway in between. These new values are used to obtain a new temperature estimate. The process is repeated until convergence is reached. It should be pointed out that although this iteration method generally gives good predictions outside the fusion zone, there is no guarantee that it will converge to the correct solution since the approximation used may not satisfy the energy conservation law. In addition, the conventional point heat source closed form solution fails to give good results in the case of multipass welding. This is due to the fact that the solution is based on the point source being located at the top surface of the plates being welded.

To accommodate the multipass welding case, a modification of the solution was made [2208] enabling one to locate the heat source at any point through the plate's thickness. It thus becomes possible to simulate each welding pass by positioning the point source at the center of the pass. The basic assumptions of the conventional solution were kept the same. Furthermore, the adiabatic boundary conditions on the top and bottom surfaces of the plate were satisfied by using the method of images. The obtained solution can then be expressed by the following equation:

$$T - T_0 = \frac{Q}{2\pi k} \cdot e^{-\lambda S_w \xi} \left\{ \frac{e^{-\lambda S_w R}}{R} + \sum_{n=1}^{\infty} \left[\frac{e^{-\lambda S_w R_n}}{R_n} + \frac{e^{-\lambda S_w R_n'}}{R_n'} \right] \right\} \quad (22.5)$$

where $R_n = \sqrt{\xi^2 + y^2 + (OT_n + z)^2}$

$R_n' = \sqrt{\xi^2 + y^2 + (OB_n + z)^2}$

$OT_n = \sqrt{OB_{n-1} + 2.F}$

$OB_n = \sqrt{OT_{n-1} + 2.G}$

with F and G being the distances of the point heat source from the top and bottom surfaces of the plate, respectively, and OT_n and OB_n the distances of the point heat source from the nth imaginary ones with respect to the top and bottom surface respectively. All other quantities have been previously defined.

22.4.1.2 *Finite Difference Solution*

Various investigators have worked on analytical models to predict the temperature distribution due to welding using various simplifying assumptions. Rosenthal [2205] did his pioneering work in this regard followed by Wells [2209], Adams [2210], Christiansen et al. [2211], Nunes [2212], Eager et. al. [2213], Goldak et al. [2214], Kasuya et al. [2215, 2216] and many others. However all these works have some or other limitation in describing the three dimensional temperature field due to a welding run. For example, the double ellipsoidal distribution of heat source is described by using five unknown parameters [2214, 2217]. Kamala et al. [2218] has suggested use of a modified power density distribution function (PDDF) for reducing errors in approximating a 3-D heat transfer problem to a 2-D one. However in this method a 3-D solution is required to modify the PDDF for the 2-D approximation. Considerable efforts have been devoted to develop numerical solutions to predict thermal responses in weldments. Although some insightful results have been obtained, most of the numerical procedures are extremely complex and difficult to implement in dealing with realistic welded structures. The thermal response of a weldment is a 3-D problem that requires considerable amount of computational time.

The heat source is modeled on the assumption that the heat flux is distributed uniformly, below the welding arc, over a rectangular parallelepiped as shown hatched in Fig. 22.6. Welding is carried out along y-axis and the origin of the coordinate system is at O as shown in Fig. 22.6. The heat lost to the environment during the instantaneous transfer of weld metal to the plate as well as the losses due to radiation and convection from the plate surface to the ambient atmosphere, conduction through the electrode were accounted for by the arc efficiency parameter η. This value was multiplied by the gross heat generated by the arc to estimate the net heat made available to the plate.

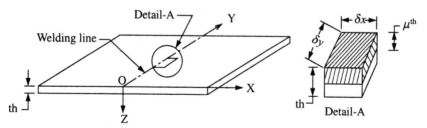

Fig. 22.6   Axis system and heat input details

Heat Flow

The governing equation of heat conduction in solid is given by

$$\rho C_p \frac{\partial T}{\partial t} = \nabla(\lambda \nabla T) + Q'_m \qquad (22.6)$$

where Q'_m = Heat input or loss per unit volume, W/m^3
   ρ = Density of metal, kg/m^3
   λ = Thermal conductivity of metal, W/mK
   C_p = Specific heat of metal, J/kg K
   T = Temperature, K

A finite difference numerical model based on a calculation procedure of simultaneous forward and backward marching within the domain is taken for solving the above mentioned heat conduction equation. By using a predictor-corrector method, temperature dependence of specific heat and thermal conductivity are incorporated. The solution of equation (22.6) gives the 3-D temperature profile due to welding.

22.4.1.3   *Finite Element Solution*
Most investigators agree that perhaps the most critical input required for the welding thermal analysis is the power, Q, that enters the plate or section being welded. It is customary to express this total heat input by expression (22.7). Of equal importance to the magnitude of the total heat input is its distribution.

$$Q = \eta VI \qquad (22.7)$$

where V = arc voltage
   I = welding current
   η = heat transfer efficiency

In the solutions presented earlier, a point or line heat source was assumed. As Rykalin [2219] and other investigators report, however, a more realistic approach is to assume a Gaussian radial heat flux distribution of the form

$$q(r) = q_0 \cdot e^{-cr^2} \qquad (22.8)$$

where q_0 = maximum heat flux at center of heat spot = cQ/π (W/mm^2)
   c = heat flux concentration coefficient = $3/r_h^2$ (mm^{-2})
   r = radial distance from center of heat spot (mm)
   r_h = radius of heat spot (mm)

Equation (22.8) is valid for a stationary arc. High-speed cinematography reveals, however, that during welding, when the arc is moving, the arc column is not radially symmetric but rather distorted backwards. This observation led Tasi [2220] to propose the following equation for the arc heat distribution instead of equation (22.8):

$$q(r, \xi) = q_0 \cdot e^{-cr^2 - \lambda S_w \xi} \qquad (22.9)$$

where all the symbols have been previously defined.

Based on the preceding discussion it is evident that in order to more accurately describe the temperature distribution and cooling rates in the region close to the weld, more realistic assumptions should be employed. A three-dimensional finite heat source model for solving the governing partial differential equation of heat transfer was thus developed [2208] under the following assumptions:

- Quasi-stationary state, that is, steady-state conditions with respect to a coordinate system moving with the heat source.
- The heat input is provided by a moving three-dimensional skewed normally distributed heat source moving on the surface of the plate and given by equation (22.9).
- The thermal conductivity k of the material is assumed to be a linear function of temperature T given by

$$k(T) = k_0 \cdot [1 + \lambda (T - T_0)]$$

where k_0 is the value of the thermal conductivity at the initial plate temperature T_0 and y the proportionality constant.
- The thermal diffusivity λ of the material is assumed to be constant.
- Convective and radiation boundary heat losses from the plate's surface are taken into account through a constant average "effective" heat-transfer coefficient h, which can be different for the top and bottom surfaces of the plate.
- The initial temperature T_0 of the plate can be different from the environmental (ambient) temperature T_e to allow for preheating.
- Phase transformation and joule heating effects can be neglected.

The finite difference method had initially the edge, the advantages of the finite-element method, especially if coupled with thermal stress analysis, are more and more recognized today. Over the years many finite-element programs have thus been developed that are capable of performing heat-transfer analyses. Several of these codes can take various nonlinearities into account in a more or less sophisticated manner. One of the most sophisticated ones is ADINAT (Automatic Dynamic Incremental Nonlinear Analysis for Temperatures) developed by Bathe and co-workers [2221] over a period of years. Some details of this program will be discussed later. It suffices here to mention that ADINAT can take into account temperature dependent material properties as well as nonlinear convection and radiation boundary conditions.

One of the significant features of the finite-element method is that it can take into account any nonlinear dependence of the material physical properties with temperature, something that is very important in the welding analysis. Furthermore, it can incorporate the latent heat of fusion or of any solid-state material transformation, which can not be furnished by closed form analytical solutions [2203, 2222].

The welding heat transfer problem is truly a three dimensional one. However if it is assumed that the welding arc is moving at a constant speed along a straight path, the speed is sufficiently high relative to the material's characteristic diffusion rate and the end effects resulting from either initiation or termination of the welding process can be neglected, the three dimensional character of the heat transfer problem can be simplified. This is achieved by analysing a cross section of the weldment of unit thickness and located in the mid-length region of the weld [2208, 2223].

The heat-transfer mechanism in the weld metal, when molten, is extremely complex. These complexities arise not only from the difficulty involved in modeling the welding arc heat flux correctly, but also from the behavior of the convective motion of the molten metal, the thermal properties of the molten metal (including the phase transformations that take place during melting and solidification), the electric heating due to the current flow in the base metal, the boundary conditions for heat losses, etc. A look at the available literature reveals that there are generally three ways for handling the problem.

The first one, still at its developing stages, tries to understand and subsequently mathematically model the physical phenomena involved, that is, the fluid flow, the convective heat transfer, etc. [2224]. Since no conclusive general results are available yet, this approach cannot be generally used.

In the second way the problem is divided into two parts. First the shape of the molten pool is semi-empirically determined, then the heat flow equations are solved numerically in the solid metal only using the melting isotherm as a boundary condition [2220]. This method cannot be used, however, because the temperature distribution in the weld metal has to be calculated too if a stress analysis is to follow (most plastic deformation that causes the formation of residual stresses takes place in this region).

The third and final approach tries to simulate the convective heat-transfer mechanism in the molten metal by using a value for the thermal conductivity of the molten metal an order of magnitude higher than that of the material at the solidus temperature. This approach, although strictly not physically correct, yields physically near correct results.

Finally, since there can be a situation of multipass welding, it is necessary to find a way to model the laying of the various beads during the welding cycle. This can be achieved by the element birth and death capabilities of the code. The program ADINAT is capable of giving birth to a predetermined number of elements at predetermined time instance, thus enabling one to model the laying of a bead by specifying the appearance of the elements representing it at the time it physically appears.

22.4.2   Finite Element Analysis of Residual Stresses

Using the temperature distributions predicted on the basis of the techniques in the previous section, one can calculate the transient strains, transient stresses and residual stresses due to welding since the problem is assumed to be uncoupled. The calculation of strains and stresses, however, poses a much more formidable problem than the one encountered in the heat-transfer analysis, making the use of numerical techniques a necessity. These difficulties stem from the complicated thermal-elastic plastic state developed in and around the weld metal during welding. Two general techniques have been developed to solve the problem. One is a simple one-dimensional analysis and the other a more sophisticated one based on the finite-element method. Both will be considered here, although emphasis will be placed on the latter.

22.4.2.1 One Dimensional Analysis

The basic assumption inherent in the one-dimensional stress analysis is that the only stress present is a function of the transverse distance from the weld centerline only. As a consequence, the equilibrium conditions are not satisfied. Despite this, however, it appears that the obtained solutions correlate reasonably well with experimental data in certain cases, as will be further explored. The algorithm for solving the problem is based on the method of successive elastic solutions as proposed by Mendelson [2225]. The program can take into account the temperature dependence of all material properties, any type of strain hardening and can solve bead-on-plate, edge, and butt welds of flat plates with finite width. One of the input requirements, the temperature distribution, can also be calculated if desired by the same program using the line heat source solution. The output at each time step consists of the temperature, total strain, mechanical strain, plastic strain, and stress at each of the predetermined points located at various transverse distances from the weld centerline.

Previous applications of the program have shown that it give good results in the case of thin plates. This happens because in this plates all stresses, except σ_x, are very small, sometimes of the order of magnitude smaller than σ_x. Finally, it should be mentioned that the one-dimensional analysis can be used with proper minor modifications for the analysis of simple structural forms other than plates.

22.4.2.2 Finite Element Formulation

The complex behavior of a weldment, and in particular the highly nonlinear material response and the material loading and unloading that occur in the multidimensional stress space, can be handled more accurately using numerical techniques such as the finite element method.

In this subsection some details on the finite element formulation of the welding stress problem are presented. Some basic considerations regarding the thermo-elastic-plastic model used are discussed in the following. In the thermo-elastic plastic model, assuming infinitesimal strains, the total strain ε_{ij} at time t is assumed to be given by

$$\varepsilon_{ij} = \varepsilon_{ij}^E + \varepsilon_{ij}^P + \varepsilon_{ij}^{TH} \qquad (22.10)$$

where ε_{ij}^E = elastic strain, ε_{ij}^P = plastic strain and ε_{ij}^{TH} = thermal strain.

Hence, at any time t during the response, the stress is given by the constitutive law for an isotropic thermo-elastic material

$$\sigma_{ij} = C_{ijrs}^E (\varepsilon_{ij} - \varepsilon_{ij}^P - \varepsilon_{ij}^{TH}) \qquad (22.11)$$

with C_{ijrs}^E denotes a component of the elastic constitutive tensor.

The thermal strains are

$$\varepsilon_{rs}^{TH} = \alpha_m (T - T_R)\delta_{rs} \qquad (22.12)$$

where α_m is the average thermal expansion coefficient T_R the reference temperature and δ_{rs} the Kronecker delta. This term can be modified to include, in addition to the thermal strain, the strains that are induced from the solid phase transformations occurring during the heating and cooling stages of the thermal history [2208, 2226].

For plastic strains ε_{rs}^P, the situation is more complicated. Although the classical theory of isothermal plasticity is a well tested one, extension of the theory to non-isothermal cases is difficult to substantiate experimentally. Relatively good agreement between theory and experiments has generally been reported but for temperatures up to about 538°C only. For lack of any alternative, however, the same non-isothermal theory of plasticity can be used throughout the temperature range encountered in welding problems.

The general form of the yield or loading function for multiaxial stress conditions is given by

$$F = F(\sigma_{ij}, \alpha_{ij}, \sigma_y) \tag{22.13}$$

where σ_{ij} and σ_y are functions of the history of plastic deformation and temperature. For elastic behavior, $F < 0$ and for plastic behavior $F = 0$. As a consequence of Drucker's postulate for stable plastic materials, F defines a convex surface in the stress-temperature space. It is also assumed that the isothermal normality condition remains valid so that

$$\dot{\varepsilon}_{rs}^p = \lambda \frac{\partial F}{\partial \sigma_{ij}} \tag{22.14}$$

where λ is a positive scalar. The selection of a hardening rule is also required for the calculation of λ. Because cyclic plasticity is expected in the welding problem, the kinematic hardening mechanism, thought by many to better model the phenomena involved, should be chosen. The assumptions involved in this mechanism are, the size of the yield surface depends on the temperature only, whereas the translation rate of the yield surface in the stress space depends on the plastic strain rate. Further details on the numerical aspects of the problem can be found in [2227].

A cross section of the weldment in its midlength can be used to calculate the transient strains, transient stresses, and residual stresses due to welding. This is rationalized by the fact that for relatively long plates the maximum stresses are developed in this region. Furthermore, the plane-strain assumption can be used (that is, all plane sections normal to the weld line remain plane during the entire welding process).

The boundary conditions to be used in the analysis should be such so as to allow free expansion of the weldment in the transverse direction as well as bending. At the same time the structure should be properly restrained to eliminate all possible modes of rigid-body motion, otherwise the stiffness matrix will not be positive definite.

The foregoing holds for a relatively thick weldment. In case of thin plate, a plane-stress analysis can alternatively be performed over the top surface of the plate (assuming negligible through-thickness stress gradients).

Material properties

The temperature dependence of young's modulus E, Possion's ratio v, yields stress σ_y, and strain hardening modulus E_T, are all required input for the thermo-elastic-plastic model used.

A word of caution is appropriate at this point as far as the material properties above the liquidus temperature are concerned. The material does not have any strength when molten since all its mechanical properties are zero. But due to numerical considerations, zero properties cannot generally be entered as input to the program. Hence to avoid any instabilities (or even divisions by zero) very small values for E, σ_y and E_T should be used above the liquidus temperature.

Another point is the accumulation of plastic strains in the regions that become molten during the welding cycle. When the temperature reaches the liquidus temperature these plastic strains are physically relieved, starting to accumulate again when the metal solidifies. The presence of nonzero material properties above the liquidus, however, would cause the plastic strains not only to continue accumulating but also to reach artificially high values owing to the very low magnitude of the mechanical properties. It is therefore necessary to impose a total relief of plastic strains when the material melts.

Solution strategy

One of the most important decisions an analyst has to make when performing a nonlinear incremental stress analysis is the solution strategy to be followed, because the accuracy and the convergence characteristics of the solution depend very much on it. This is especially true for complex situations involving highly nonlinear material behavior like the one encountered in the welding problem. It is therefore necessary to perform several numerical experiments prior to a full analysis of the problem.

22.4.3 Analytical Solutions of Residual Stresses

The weld induced residual tension can be estimated from simple empirical formulae for all types of common welds and processes, with an accuracy sufficient for practical purposes [2228]. The resulting residual stress pattern can then be calculated using the principles of static equilibrium.

Weld shrinkage can also be predicted, although perhaps not with the same degree of confidence. The reason for this uncertainty is however not attributable to any shortcoming in mathematical modeling, but the difficulty of defining realistic boundary conditions as well as other arbitrary and variable effects such as weld penetration, heat losses etc. It is therefore unlikely that any improvement would accrue from more complex or rigorous analysis without disproportionate trouble and cost.

Rigorous theoretical determination of residual stresses and thermal distortion is complicated by the creation of localised transient zones of plasticity as the heat source moves along the workpiece. Further difficulties arise in accounting for heat losses due to convection and radiation from the heated areas, factors which add a random element to any method of predicting thermal stresses. The physical basis of thermal distortion and the factors affecting it are not difficult to understand, and therefore empirically determined formulae are adequate for practical purposes.

this has been formulated by White [2231]. The process is iterative but simple and rapidly convergent:

(a) Guess a value for the residual stress σ_r
(b) Calculate A_t from equations (22.16) and (22.17) for a single pass (simultaneous welds are additive)
(c) Estimate the total tendon force for a butt weld from the formula

$$F = (\sigma_y + \sigma_r)\left(2c + \frac{f + g}{2}\right)t_h \qquad (22.22)$$

where $f + g$ = top + bottom widths of V edge preparation,

$$c = \sqrt{2A_t / \pi}$$

or for a T-fillet weld from the formula

$$F = (\sigma_y + \sigma_r)\left[t_h(c + 1/2a) + c(t_h + a) + 1/2\pi c^2\right] \qquad (22.23)$$

where a = leg length

$$c = \sqrt{A_t(1 + \pi/2)} \quad \text{if} \quad A_t < (1 + \pi/2)t_h^2$$

$$c = (\sqrt{t_h^2 + 2\pi A_t} - t_h)/\pi \quad \text{if} \quad A_t > (1 + \pi/2)t_h^2$$

t_h = web plate thickness
(note that c must be less than the flange plate thickness)

(d) Calculate σ_r from equation (22.15) and if significantly different from the assumed value, repeat (b) and (c).

The total area of the weld affected zone is obtained from equation (22.16) using the final calculated tendon force which is assumed to act at the center of area for the purpose of estimating the bending stress (Fig. 22.8).

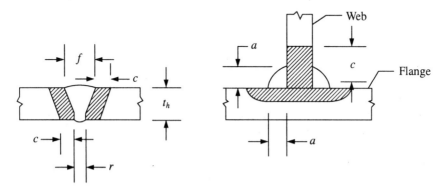

Fig. 22.8   Definition of terms in equations 22.22 and 22.23

REFERENCES

2201 Masubuchi, K., *Analysis of Welded Structures: Residual Stresses, Distortion and their Consequences*, Pergamon Press, Oxford/New York, 1980.

2202 Mroz, Z. and Raniecki, B., "On the uniqueness problem in coupled thermoplasticity" & "A derivation of the uniqueness condition in coupled thermoplasticity", *International Journal of Engineering Sciences*, Vol. 14, 1976, pp. 211–221 & 395–401.

2203 Hibbitt, H.D. and Marcal, P.V., "A numerical thermomechanical model for the welding and subsequent loading of a fabricated structure', *Computers and Structures*, Vol.3, 1973, pp. 1145–1174.

2204 Ueda, Y., Fukuda, K. and Nakacho, K., "Basic procedures in analysis and measurement of welding residual stresses by the finite element method", *Proceedings, International Conference on Residual Stresses in Welded Construction and their Effects*, The Welding Institute, London, 1977, pp. 27–37.

2205 Rosenthal, D., "Mathematical theory of heat distribution during welding and cutting", Welding Journal, Vol. 20, No. 5, 1941, pp. 1277–1292.

2206 Rosenthal, D., "The theory of moving sources of heat and its application to metal treatment", *Transactions, ASME*, Now. 1946, pp. 849–866.

2207 Myers, P.S., Uyehara, O.A. and Borman, G.L., "Fundamentals of heat flow in welding", *Welding Research Council Bulletin*, No. 123, July, 1967.

2208 Papazoglou, V.J., "Analytical techniques for determining temperatures, thermal strains and residual stresses during welding", *Ph.D. Thesis*, Department of Ocean Engineering, M.I.T., Cambridge, Mass., May 1981.

2209 Wells, A., 1952, *Heat flow in welding*. Welding Journal 31(5), pp. 263–267.

2210 Adams, C.M., Jr. 1958, *Cooling rates and peak temperatures in fusion welding*. Welding journal 35(5), pp. 210–215.

2211 Christiansen, N., Davies, V., Gjermundsen, K., 1965, *Distribution of temperatures in arc welding*. British Welding Journal, 12(2), pp. 54–75.

2212 Nunes, A.C., Jr., 1983, *An extended Rosenthal weld model*. Welding Journal, 62(6), pp.165–170.

2213 Eager, T. and Tsai, N., 1983, *Temperature fields produced by travelling distributed heat sources*. Welding Journal 62(12), pp. 346–355.

2214 Goldak, J., Chakravarti, A. and Bibby, M., 1984, *A new finite element model for welding heat sources*. Metallurgical Transactions, 158(2), pp. 299–305.

2215 Kasuya, T. and Yurioka, N., 1989, *Analysis of three-dimensional heat conduction in welding by dispersed point heat sources*. IIW doc. IX, 155–89.

2216 Kasuya, T. and Yurioka, N., Mar. 1993, *Prediction of welding thermal history by a comprehensive solution*. Welding Journal, pp.107–115.

2217 Nguyen, N.T., Ohta, A., Matsuoka, K., Suzuki, N., Maeda, Y., Aug. 1999, *Analytical Solutions for Transient Temperature of Semi- Infinite Body subjected to 3–D Moving Heat Sources*, Welding Journal, pp. 265–274.

2218 Kamala, V. and Goldak, J.A., Sept. 1993, *Error due to two dimensional approximation in heat transfer analysis of welds. Welding Journal*, pp. 440–446.

2219 Rykalin, N.N. and Nikolaev, A.V., "Welding arc heat flow", *Welding in the World*, Vol. 9, No. 3/4, 1971, pp. 112–132.

2220 Tsai, C.L., "Parametric study on cooling phenomena in underwater welding", *Ph.D. thesis*, Department of Ocean Engineering, MIT, Cambridge, Mass., Sept. 1977.

2221 Bathe, K.J. and Khoshgoftaar, M.R., "Finite element formulation and solution of nonlinear heat transfer", *Journal of Nuclear engineering and Design*, Vol. 51, 1979, pp. 389–401.

2222 Friedman, E., "Thermomecanical analysis of welding process using the finite element method", *Journal of Pressure Vessel Technology*, ASME, Aug. 1975, pp. 206–213.

2223 Friedman, E., "Numerical simulation of the gas tungsten arc welding process", Proceedings, *Numerical Modeling of Manufacturing Processes*, ASME Winter Annual Meeting, Atlanta, Ga., 1977, pp. 35–47.

2224 Dilawari, A.H., Szekely, J. and Eagar, T.W., "Electromagnetically and thermally driven flow phenomena in electroslag welding", *Metalurgical Transaetions* B, Vol. 9B, 1978, pp. 371–381.

2225 Mendelson, A., *Plasticity: Theory and Application*, Macmillan, New York, 1968.

2226 Papazoglou, V.J. and Masubuchi, K., "Numerical analysis of Thermal Stresses During Welding Including Phase Transformation Effects", *Journal of Pressure Vessel Technology*, ASME, Vol. 104, No. 3, Aug. 1982, pp. 198–203.

2227 Synder, M.D., "An effective solution algorithm for finite element thermo-elastic-plastic and creep analysis", *Ph.D. Thesis*, Department of Ocean Engineering, MIT, Cambridge, Mass., Oct. 1980.

2228 Rectliffe, M.A., "The basis and essentials of thermal residual distortion in steel structures", *The Royal Institution of Naval Architects*, 1983.

2229 White, J.D., "Longitudinal stresses in a member containing non-interacting welds", *CUED Report, C-Struct/TR.58*, 1977.

2230 Martyr, D.R., "High power laser applications in the shipbuilding industry", *Trans. NECIES*, Vol. 99, 1983, pp. 106–109.

2231 White, J.D., "Longitudinal shrinkage of multi-pass welds", *CUED Report, C-Struct/TR59*, 1977.

Index